D1530047

Environmental NGOs in World Politics

Linking the local and the global

Thomas Princen and Matthias Finger

With contributions by
Jack P. Manno and Margaret L. Clark

BOWLING GREEN STATE UNIVERSITY
DISCARDED
LIBRARY

London and New York

BOWLING GREEN STATE
UNIVERSITY LIBRARIES

First published 1994
by Routledge
11 New Fetter Lane, London EC4P 4EE

Simultaneously published in the USA and Canada
by Routledge
29 West 35th Street, New York, NY 10001

© 1994 Thomas Princen and Matthias Finger

Typeset in Times by
Ponting–Green Publishing Services, Chesham, Bucks

Printed and bound in Great Britain by
T.J. Press (Padstow) Ltd, Padstow, Cornwall.

All rights reserved. No part of this book may be reprinted or
reproduced or utilized in any form or by any electronic,
mechanical, or other means, now known or hereafter
invented, including photocopying and recording, or in any
information storage or retrieval system, without permission in
writing from the publishers.

British Library Cataloguing in Publication Data
A catalogue record for this book is available from the British
Library.

*Library of Congress Cataloging in Publication Data has been
applied for.*

ISBN 0–415–11509–4 (hbk)
ISBN 0–415–11510–8 (pbk)

To Carmencita and Andrea

Contents

Preface

Three years ago we formed a faculty seminar on international environmental politics at Syracuse University's Program on the Analysis and Resolution of Conflicts and at SUNY's College of Environmental Science and Forestry. Our purpose was initially simple: to identify and understand key actors and processes associated with efforts to reduce or reverse current trends in global environmental degradation. We searched for well-documented empirical studies and useful theories, but regularly came up short.

Along the way, we came upon Lynton Caldwell's writings and, in particular, his recognition of a growing phenomenon, namely, the rise in numbers and activities of international environmental non-governmental organizations (NGOs). He wrote in 1988 that NGO action has been

> absolutely essential to most international environmental action . . . [and] much less visible than action by the national and intergovernmental bureaucracies that actually administer international environmental programs. The nature and extent of NGO influence on international environmental policy has not received comprehensive or detailed study.
>
> (Lynton K. Caldwell, 'Beyond Environmental Diplomacy: The Changing Institutional Structure of International Cooperation', in John E. Carroll, ed., *International Environmental Diplomacy* (Cambridge: Cambridge University Press, 1988) 24.

As we began to warm to the challenge, we probably gave insufficient attention to Caldwell's warning for those who might attempt such work: 'The cost of such study would be considerable and is not likely to be borne by any of the conventional sources of research funding.' Indeed, conventional sources did not support this research. A number of less conventional ones did, however, including the Canadian Consulate, Syracuse and the University of Michigan. For this, we are most grateful.

But the costs Caldwell refers to, as we discovered, were not just financial. They were costs associated with the difficulty of documenting and conceptualizing such a slippery phenomenon as NGO relations. We began to envy those who restricted their inquiries in the field of international environmental politics to the state system and treaty writing or to comparative national policies or even to the politics of scientific communities. We became rather jealous of those who could operate from a well established conceptual framework such as social movement theory, international political economy, economic development, or regime and cooperation theory.

With no 'theory' of world environmental politics generally, let alone the data necessary to conceptualize the NGO phenomenon, we began with an empirical focus. We dug up whatever we could that documented the NGO role. We found many references to the importance of NGOs and anecdotal descriptions of their work, but we found very little that revealed the details of NGO interactions in a given decision-making situation. We found that, whereas one could find an abundance of books documenting environmental conditions and many prescribing remedies to save the planet, there was precious little on the details of what, exactly, key actors, including NGOs, were doing. Moreover, we found that, although much has been written about NGOs from a social action or social movement perspective, there has been little conceptualization of the NGO phenomenon as a political development in its own right. To understand what NGOs actually did in world politics, we realized that we would have to incur the costs of doing original case studies and of conceptualizing the NGO role.

We began this study, therefore, with modest ambitions, given the poor empirical and theoretical state of affairs. We accepted Caldwell's challenge insofar as we would generate the beginnings of a useful data base and venture some preliminary propositions regarding the role of environmental NGOs in world politics. After all, it was only to be a year-long project, as both of us were on short-term visiting appointments.

Three years and several geographical and career moves later, we find it hard to stop. Almost daily we discover new bits and pieces to fill out the NGO picture. And more and more of our colleagues are acknowledging the importance of NGOs, and some are even studying their role. More important, almost daily we reconceptualize the NGO role. Even if we had wanted to fit the NGO phenomenon into conventional categories of green parties, or public interest groups, or whatever, we found we could not. We found that the more we take account of biophysical and social conditions relating to global environmental degradation, the more we must conceptualize the NGO role *de nouveau*.

As a result, we increasingly view the NGO phenomenon in world politics as critical, fluid, and, possibly, ephemeral. We see NGO politics as a crucial counterweight to dominant trends in the global political economy and at all levels, from the local to the global. We see NGO activity as essential to societies' movement toward forms of governance consistent with sustainability. We do not see NGOs, however, as replacements for other actors, namely governments and businesses. NGOs are critical because the biophysical and social conditions necessary for sustainability must be translated into a politics that is at once local and global, and both economic and moral.

This translation is not being made by the dominant actors, states and corporations. If NGOs succeed, however, they will work themselves out of a job, at least the job they perform today.

This study has benefited from the advice, encouragement, and criticism of many individuals. In the original seminar, Louis Kriesberg, Margaret Shannon, Errol Meidinger, and Stuart Thorson were regular and valued contributors to the early inquiry. In this seminar, we were fortunate to attract and, later, have write with us, one of those rare individuals who can cross the worlds of academe and practice and who can push people on both sides to examine their assumptions and play out their logics. Jack Manno has kept us on the ground with his in-depth understanding of grass-roots and international environmental politics and his knowledge of the biophysical and social conditions underlying those politics. Manno's Great Lakes case study and his contribution to the concluding chapter and the volume as a whole have been invaluable.

In a conference we held in Ann Arbor in October 1991, Margaret Clark joined our inquiry and contributed the case study on Antarctica. Lynton Caldwell also participated in the conference, challenging us and inspiring us to continue. Many others contributed their thoughts and insights then and in subsequent meetings of the International Studies Association's Environmental Studies Section, as well as in discussion and correspondence. Among them are: Marie Balle, Marie Lynn Becker, Mimi Becker, Garry Brewer, Fred Brown, Bunyan Bryant, James Crowfoot, Simon Dalby, Kristin Dawkins, Elizabeth Economy, Tim Eder, Kent Fuller, Michael Gilbertson, John Hough, John Jackson, Sally Lerner, Ronnie Lipschutz, Anthony Lyon, André McCloskey, Anne Marie McShea, Marie Lynn Miranda, Gail Osherenko, Elizabeth Owen, Henry Regier, Michael Ross, Paul Sampson, Wayne Schmidt, Steven Schneider, Andrew Schultheiss, Jennifer Sell, Ronald Shimizu, Richard Smardon, John Soluri, Detlef Sprinz, Cindy Squillace, Ted Trzyna, Richard Tucker, Mark Van Putton, Konrad von Moltke, Paul Wapner, Wendy Woods, Steven Yaffee, and Oran Young. We would also like to thank three anonymous reviewers for their helpful suggestions and, for copy editing and preparing the manuscript, Virginia Barker, Kathy Hall and Laura Frank.

Thomas Princen, Ann Arbor, Michigan, USA
Matthias Finger, New York, New York, USA
November 1993

List of acronyms used

AECA	African Elephant Conservation Act of 1988
AECCG	African Elephant Conservation Coordinating Group
AERSG	African Elephant and Rhino Specialist Group
AMIC	Australian Mining Industry Council
ANEN	African NGOs Environment Network
ANGOC	Asian NGO Coalition
ASOC	Antarctic and Southern Ocean Coalition
ATCP	Antarctic Treaties Consultative or Contracting Parties
ATS	Antarctic Treaty System
BCSD	Business Council for Sustainable Development
BIOMASS	Biological Investigations of Marine Antarctic Systems and Stocks
BNGOF	Brazilian NGO Forum
BRAC	Bangladesh Rural Advancement Committee
CAN	Climate Action Network
CAPE '92	Coalition to Protect the Earth 1992
CCAMLR	Convention on the Conservation of Antarctic Marine Living Resources
CF	US Conservation Foundation
CFC	Chlorofluorocarbon
CHM	Common Heritage of Mankind
CIS	Commonwealth of Independent States
CITES	Convention on International Trade in Endangered Species of Wild Fauna and Flora
COCF	Center for Our Common Future
CONGO	Conference of Non-Governmental Organizations
CRAMRA	Convention on the Regulation of Antarctic Mineral Resources Activities
CSAGI	Comité Special de l'Année Géophysique Internationale
CUSIS	Canada-United States Inter-University Seminar
ECOSOC	United Nations Economic and Social Council
EDF	Environmental Defense Fund
EEB	European Environmental Bureau
EEC	European Economic Community

EIA	Environmental Investigation Agency
ELCI	Environmental Liaison Center International
ENDA	Environment Development Action in the Third World
EPA	US Environmental Protection Agency
FAO	United Nations Food and Agricultural Organization
FoE	Friends of the Earth
FWS	US Fish and Wildlife Service
GAO	US Government Accounting Office
GATT	General Agreement on Trade and Tariffs
GEF	Global Environmental Facility
GLU	Great Lakes United
GLWQA	Great Lakes Water Quality Agreement
GTC	Global Tomorrow Coalition
ICC	International Chamber of Commerce
ICC	Inuit Circumpolar Conference
ICSU	International Council of Scientific Unions
IEB	International Environmental Bureau
IFC	International Facilitating Committee
IFYGL	International Field Year on the Great Lakes
IGO	Intergovernmental organization
IIED	International Institute for Environment and Development
IJC	International Joint Commission
IOPN	International Office for the Protection of Nature
ITRG	Ivory Trade Review Group
ITTO	International Tropical Timber Organization
IUCN	International Union for the Conservation of Nature (World Conservation Union)
JATAN	Japan Tropical Forest Action Network
MUCC	Michigan United Conservation Clubs
NCP	Non-consultative parties
NGO	Non-governmental organization
NRDC	Natural Resources Defense Council
NWF	National Wildlife Federation
PAN	Pesticides Action Network
PLUARG	Pollution from Land-Use Activities Reference Group
PrepCom	Preparatory Committee
RAP	Remedial Action Plan
RSCNRC	Royal Society of Canada and the National Research Council
SCAR	Scientific Committee on Antarctic Research
SCM	Special Consultative Meeting
TRAFFIC	Trade Records Analysis of Flora and Fauna in Commerce
UAE	United Arab Emirates
UN	United Nations
UNAEPA	United Nations Antarctic Environmental Protection Agency
UNCED	United Nations Conference on Environment and Development

UNEP	United Nations Environment Programme
UNESCO	United Nations Educational, Scientific, and Cultural Organization
UNGA	United Nations General Assembly
WICEM I	First World Industry Conference on Environmental Management
WICEM II	Second World Industry Conference on Environmental Management
WRI	World Resources Institute
WTMU	Wildlife Trade Monitoring Unit
WWF-Int.	Worldwide Fund for Nature-International
WWF-US	US World Wildlife Fund

1   Introduction

Thomas Princen and Matthias Finger

In the fast-growing literature on international environmental affairs, two phenomena regarding environmental non-governmental organizations (NGOs) stand out. One is the tremendous growth in the size and numbers of environmental NGOs. The second, with a sizeable yet understandable lag, is the growing awareness among scholars that this phenomenon is not 'epi-phenomenal', but integral to the peculiar nature of world environmental politics itself.[1] The role of NGOs in the international arena is not strictly analogous to the role of groups who lobby and raise public awareness in the domestic arena. Nor is their role to replace governments. At the international level, environmental NGOs do lobby and educate and substitute for govern-ments, but their peculiar contribution is something quite different as well. Our task in this book, then, is to characterize the distinctive qualities of NGO relations and, hence, a distinctive feature of world environmental politics.[2] As will be seen, our focus on NGO *relations* draws analytic attention to processes (not just to international structures of power and institutions), to strategic interactions (not just to education), and to the transformative effects of NGO activity in the world political economy (not just to the ameliorative and reactive functions of NGOs).

The sheer numbers of NGOs worldwide, let alone the size and scope of some individual NGOs, are striking. Possibly most significant is the growth in these numbers this century and, especially, just since 1980. Some data are illustrative.[3]

International organizations generally have grown rapidly this century. But whereas between 1909 and 1988, intergovernmental organizations grew from thirty-seven to 309, *non*-governmental organizations grew from 176 to 4,518.[4] Thus, the increase in NGOs can be explained only in part by the proliferation of international organizations generally. Comparable data on international *environmental* NGOs are not available, but indirect indicators suggest that their growth has been at least as dramatic as that of international NGOs generally. In fact, almost all environmental NGOs, networks, and coalitions were started in the 1980s.[5]

Membership in international NGO coordinating bodies is one indicator of NGO growth. The Environmental Liaison Center International (ELCI), the

NGO liaison unit with the United Nations Environment Programme (UNEP), had 726 member organizations in 1993, a figure which ELCI says has been steadily increasing since its creation in 1972.[6] The World Conservation Union (IUCN) lists its NGO membership at 450.[7] Twenty-one African NGOs formed the African NGOs Environment Network (ANEN) in 1982. This number increased more than ten-fold in its first six years and, by 1990, the membership was 530 NGOs, located in 45 countries.[8]

In-country numbers are also impressive. One study estimates that there are more than 6,000 NGOs in Latin America and the Caribbean, most of these formed since the mid-1970s.[9] In Brazil, for example, there were 400 NGOs in 1985 and 1,300 in 1991.[10] And a survey of 1,000 NGOs in Brazil found that 90 per cent were started since 1970.[11] In Kenya there are some four hundred to six hundred NGOs, of which more than one hundred are international in their operations. Asian countries probably have the largest number of NGOs in the developing world.[12] In Indonesia, for example, WALHI, the Indonesian Environmental Forum, was formed by seventy-nine NGOs in 1980, had grown to over 320 NGOs by 1983 and, in 1992, had over 500 members.[13] India has some 12,000 development NGOs and probably hundreds of thousands of local groups.[14] Bangladesh has more than 10,000 environment-related NGOs, of which about 250 receive funds from foreign sources.[15] The Philippines has some 18,000 NGOs, mostly rural and small, but some internationally prominent. In the former Soviet Union, one study listed 331 environmental groups in 1990 during the *glasnost* period, of which 235 were in the Russian Federation and 52 in the Ukraine.[16]

Another indicator of the growing numbers and prominence of NGOs worldwide is the number of directories that have sprung up in recent years. The *World Directory of Environmental Organizations,* now in its fourth edition, lists 365 international environmental NGOs in just one chapter. The *International Directory of Non-Governmental Organizations* lists some 1,650 environmental and development NGOs interested in multilateral development bank issues.[17] The *Who is Who in Service to the Earth* of 1991 lists about 2,500 organizations, many of which are environmental.[18]

Yet another indicator of growing NGO prominence is the organizational growth which many individual NGOs, especially some of the more prominent Northern[19] groups, have experienced since the early 1980s.[20] From 1983 to 1991, for example, the revenues for the US branch of the World Wildlife Fund (WWF-US) increased from $9 million to $53 million, and its membership rose from 94,000 to more than one million. In the 1980s, WWF-US contributed $62.5 million to more than 2,000 projects worldwide.[21] From 1985 to 1990, membership in Greenpeace increased from 1.4 million to 6.75 million and annual revenues went from $24 million to some $100 million.[22] Greenpeace had five foreign affiliates in 1979, but in 1992 had offices in twenty-four countries worldwide.[23] Friends of the Earth (FoE) began as a strictly United States organization, opening its first office in San Francisco in 1969, but soon expanded to Paris (1970) and London (1971). In the early

1970s, FoE began developing an international structure called Friends of the Earth International, which grew from twenty-five member groups worldwide in 1981 to fifty-one in 1992.[24] The Nature Conservancy, founded in 1951, began its international programmes in 1974 but it was not until 1987 that a splinter group formed Conservation International; by 1991 it had twenty NGO partners in sixteen Latin American countries, and a budget of $10.9 million.[25] The Sierra Club increased its membership from 346,000 in 1983 to 560,000 in 1990 and has an annual budget of $35 million. The Natural Resources Defense Council (NRDC), founded in 1972 with 6,000 members, now has 170,000 and an annual budget of $16 million.[26] Both the Sierra Club and the NRDC expanded their international programmes in the 1980s and early 1990s.[27]

The emergence of large-scale international NGO coalitions is also striking. In Asia, the Asian NGO Coalition for Agrarian Reform and Rural Development facilitates dialogue among South and Southeast Asian NGOs and between these NGOs and Northern NGOs.[28] The International NGO Forum on Indonesia is composed of NGOs from Indonesia, the Philippines, Thailand and other parts of Asia and from such Northern countries as the Netherlands, Belgium, Germany, and the United States. The Forum has met since 1985 on an annual basis in conjunction with the meetings of the international donor aid consortium in the Hague which is responsible for foreign assistance to Indonesia.[29] In Japan, the Japan Tropical Forest Action Network (JATAN) was founded in 1987 by ten Japanese NGOs but now has a network spanning much of Asia, North America, Latin America, and Europe.[30]

In Africa, ENDA Tiers-Monde (Environment and Development in the Third World), operates mostly in West Africa but has networks throughout the continent and branches in Latin America, the Caribbean, India and the Indian Ocean. Founded in 1972 with the support of UNEP, it is now funded by a consortium of European governments. With a permanent staff of some 400 people, its work on human rights, environment and democracy has quadrupled from the early 1980s to the early 1990s.[31]

Among indigenous peoples, the Inuit Circumpolar Conference (ICC) represents indigenous peoples from the Arctic region. The ICC operates transnationally to oppose militarization, to protect cultural values and native lands, and to promote self-government.[32] The Coordinating Council of Indigenous Nations of the Amazon Basin, the Indigenous Women's Network, the World Council of Indigenous Peoples and others coordinate indigenous rights issues that span state boundaries.[33] In preparation for UNCED, forest-dwelling communities from Asia, Africa, and Latin America formed the World Alliance of the Indigenous-Tribal Peoples of the Tropical Forests. Thirty representatives of these communities drew up a charter in February of 1992 calling for the 'recognition, definition and demarcation of our territories in accordance with our local and customary systems of ownership and use' and insisted upon an end to imposed development.[34]

In the Middle East, an environmental movement has been hindered by political turmoil as well as a lack of a tradition of private support (except for nature protection in Israel, which dates back at least to the early 1950s). Nevertheless, environmental NGOs did begin to emerge in the 1980s. And although most groups have operated locally, two regional NGO networks have formed, in part to provide a politically neutral ground for coordinating action across those states bordering the Mediterranean and the Gulf of Aqaba.[35]

In Western Europe, the European Environmental Bureau (EEB) had, in 1991, 126 environmental NGOs from twenty-one European countries. The EEB focuses on environmental provisions in the European Community, has direct access to the European Commission, and represents European NGOs in many international fora.[36] In Central and Eastern Europe, a nascent coalition is organizing to monitor western business investment and to coordinate with western NGOs who have experience campaigning against such firms.[37]

In North America, Great Lakes United encompasses environmental, sporting, trade union, indigenous peoples and municipality interests to represent water quality issues in the Great Lakes basin (see Chapter 4). In the United States, the Southwest Network for Environmental and Economic Justice has some sixty affiliates dealing with issues along the US–Mexico border.[38] In their third annual meeting in San Diego, California, in August 1993, the network included environmental and social organizations from Mexico and Asia. The Global Tomorrow Coalition has 120 members, including both mainstream and grass-roots organizations, educational institutions, and corporations. Their aim is to promote sustainable development both in the US and abroad and they have increasingly attempted to involve Southern NGOs.[39] The Antarctic and Southern Ocean Coalition (ASOC), based in Washington, DC, has some 175 NGO members from thirty-three countries (see Chapter 6). In 1989, sixty-three NGOs from twenty-two countries formed the Climate Action Network.[40]

Perhaps the most telling indicator of NGOs' prominence in world politics is their increasing presence in international conferences. Since the inception of the United Nations, a pattern of parallel NGO conferences has emerged. The most prominent have been those associated with the 1972 Stockholm and 1992 Rio conferences on the environment and development (see Chapter 7). In Geneva, at the preparatory negotiations to the Rio conference, the United Nations Conference on Environment and Development (UNCED), some 300 NGOs from around the world attended, and in New York more than 1,000 attended. Some 22,000 NGO representatives of more than 9,000 NGOs then travelled to the conference itself in Rio. In New York, fifteen countries had NGO observers on their delegations, including twenty-four representatives on the US delegation.[41] At Rio, by one count, some 150 official delegations had NGO representatives.[42]

NGOs have been active in the follow-up to Rio. They participated in the formation and first session of the Commission on Sustainable Development, the only institutional innovation coming out of Rio.[43] In compliance with the Agenda 21 mandate, the UN hosted a conference in July of 1993 on migratory fish which was attended by 105 government delegations, sixteen international agencies, and forty-one NGOs.[44] At the end of the conference more than 120 Northern and Southern NGOs endorsed a statement calling for a precautionary approach to fishery management and stronger international enforcement. They also planned to strengthen their own North–South ties and work together to draft negotiating text for subsequent conferences. As a result of the unprecedented numbers and roles played by NGOs in UNCED, the United Nations' Economic and Social Council's Committee on NGOs has recommended a two-year study on the NGO relationship with the UN.[45]

Although parallel conferences are important, possibly more significant are NGO activities aimed directly at shaping international laws and institutions.[46] For example, major international NGOs such as the London-based International Institute for Environment and Development (IIED) have been involved with the International Tropical Timber Organization (ITTO) since its inception. Promoting the creation of the organization, then monitoring and doing reports for it, and, finally, decrying the lack of progress in achieving its conservation mandate, the IIED, along with Greenpeace, Worldwide Fund for Nature-International (WWF-Int.) and others, have maintained a regular presence at biannual meetings and special committee meetings. Through publicity and independent reports these groups are widely acknowledged for putting pressure on the parties to implement the conservation features of the International Tropical Timber Agreement. In fact, in the ITTO's 1990 Action Plan, NGOs are frequently cited as key actors for implementing what is now considered the primary goal of the ITTO, namely, sustainable use and ecosystem integrity.[47] And in the 1993 renegotiation of the original agreement, NGOs have joined with producing countries to expand the scope of the regime to all timber – tropical and temperate – thus forcing Northern consuming states to consider whether to apply the same standards for their forestry practices as they are promoting for those of Southern producing states.[48]

The International Whaling Commission, although for many years resistant to public participation in its meetings, has allowed increased NGO involvement. In a ten-year period, the numbers of NGOs has risen from five to fifty. These NGOs circulate information on infractions by member states and provide scientific and legal interpretations. Moreover, they have worked outside the meetings to get non-whaling states to join such that in 1982, with an expanded membership, a majority favoured a whaling moratorium.[49]

The London Dumping Convention has granted observer status to NGOs since the early 1980s. Greenpeace and other NGOs concerned with marine environments have participated actively and, in fact, have been invited to contribute their specialized skills in scientific working groups.[50]

NGOs have been widely credited with performing an instrumental role in pushing for and then strengthening the 1987 Montreal Protocol on Substances that Deplete the Ozone Layer. Owing largely to UNEP policy to involve non-state actors, NGOs participated directly in the preparatory and actual negotiations. But possibly of equal or greater impact, NGOs exerted their influence by targeting key industries. For example, after several years of direct action against British chlorofluorocarbon (CFC) producers, Friends of the Earth-UK announced in February 1988 a boycott of CFC-based aerosol products. Three days before the boycott was to begin, the industry announced a phase-out in advance of the Protocol's schedule.[51]

With respect to the multilateral banks, NGOs are often credited with putting the environment on the international development agenda. For example, at the 1991 annual meeting of the Asian Development Bank, twenty-four NGOs from Asia, the South Pacific, and North America attended and were accorded office space and access to bank officials.[52] World Bank meetings have been a focal point of NGO activity since at least 1983 when six large US NGOs pressured the Bank to include environmental costs in its projects. Largely as a result of these efforts, the Bank has added an environment department and the Global Environment Facility (GEF). Moreover, whereas NGOs were involved in about thirteen projects annually between 1973 and 1988, in 1989 they were involved in forty-six and in 1990 fifty, nearly a quarter of all approved projects that year.[53] The World Bank now meets regularly with environmental NGOs, albeit primarily with large, Washington, DC-based groups.[54]

Other international organizations have not been as accommodating of NGOs. The Food and Agriculture Organization (FAO), for example, does not provide an NGO room to facilitate NGO participation. After pressure from NGOs, especially the Pesticide Action Network, however, the FAO has allowed both industry and NGO access to an experts' working group to design a code of conduct regarding potentially toxic chemicals.[55]

Whatever the precise numbers of NGOs and their specific roles in various forums, it is clear an NGO phenomenon exists. It is less clear, however, what entities constitute that phenomenon. The term 'NGO' has many uses and many connotations. The difficulty of characterizing the entire phenomenon results in large part from the tremendous diversity found in the global NGO community. That diversity derives from differences in size, duration, range and scope of activities, ideology, cultural background, organizational culture, and legal status.

NGOs vary considerably in terms of the size of their budgets, staff, and offices. As noted, many Northern groups have multimillion dollar operations and thousands of staff spread around the world. One of the largest NGOs in the world is in the South, however: the Bangladesh Rural Advancement Committee (BRAC), with a staff of 2,000.[56] Many more NGOs, however, operate with only a few paid staff and with very limited resources. Even in the North, small size, even among those with international operations, is probably the norm. JATAN in Japan, for example, despite its global network and prominence both at home and abroad, has only four staff members and an annual budget of $191,000.[57] ASOC had only one staff member for much of its campaign to make Antarctica a world park (see Chapter 6).

NGOs vary by their organizational duration. In the United States, the Sierra Club and Audubon Society have been around for nearly a century. In Europe, the same is true for nature protection and so-called nature-friends' (*Naturfreunde*) organizations. Those who formed in the 1970s, such as IIED, Greenpeace, FoE, and NRDC also have acquired a certain degree of stability owing to strong, well-established constituencies in their respective countries. Other NGOs, especially those who form around specific incidents or events such as the Rio UNCED conference appear more fragile. The durability of NGOs is probably a function of the salience of the targeted issues as well as an NGO's capacity to organize, raise funds, and integrate into larger networks and institutions. The transient quality of NGOs pertains mostly to local NGOs, however. It is probably fair to say that NGOs which have achieved international status have, at the same time, achieved a degree of stability owing to the prominence afforded them by international networks.[58]

The range of international NGO activities varies considerably. Some NGOs seem to be everywhere. Greenpeace chases whalers in every ocean, as well as ships carrying plutonium from Europe to Japan. Other NGOs work across borders but on a highly regionalized, even localized, scale. Great Lakes United (GLU) crosses the US–Canada border, but confines its operations to the Great Lakes basin (see Chapter 4). The Coalition for Justice in the Maquiladoras has over eighty NGO members on both sides of the US–Mexico border and deals with issues local to Mexican and US communities affected by border industries.[59] The scope of activities ranges from wildlife conservation to pollution abatement to source reduction to poverty alleviation to human rights, and from research to education to lobbying to project implementation. One of the messages coming out of the 1992 Global Forum, the NGO event in Rio parallel to UNCED, was that when environment and development are brought together, NGOs of nearly all stripes emerge. A second message is that although each NGO has its special area of focus, few can achieve their aims without incorporating a wide range of activities.

NGOs also vary in their ideological orientations. In Europe, especially in Germany, the controversies between so-called 'realist' and 'fundamentalist' greens highlights the ideological differences among the various factions of the green movement. One faction is more compromising with the political system, while the other is politically more radical. Other ideological differences within the NGO community are inspired by feminism, deep ecology, spiritual ecology, social ecology and bioregionalism.[60]

Cultural differences distinguish NGOs as well. Many southern NGOs trace their roots to political and human rights challenges. In the Philippines, for example, the rapid rise in NGOs and their increasing prominence at all levels of Philippine society derive largely from the overturning of the Marcos regime and from the encouragement given by the Aquino administration to non-governmental activism.[61] In Latin America, much of the NGO activity grew out of the work of the Catholic Church in the 1970s and, especially, 'Vatican II', which called for greater social justice. In the 1980s, a broader set of interests including environmental and public health concerns have stimulated the formation of NGOs.[62] In post-colonial African countries, local NGOs have been involved in environmental, development, and basic services delivery, largely as a result of the inability of governments to provide such services. Throughout the South, women have been ignored in the development process and, partly as a result, many NGOs exclusively for women have been started.[63] In Europe, many prominent NGOs such as Greenpeace or Friends of the Earth trace their roots to the anti-nuclear movement of the 1970s. One result that carries over into the environmental realm is a penchant for scientific and technological solutions.

Differences in organizational culture set NGOs apart as well. Many Northern NGOs have undergone processes of considerable institutionalization and bureaucratization. They have developed organizational structures comparable to business organizations, with corresponding marketing, fund-raising and development departments. The headquarters of Washington-, Tokyo-, or Brussels-based NGOs look more like corporate headquarters than the grass-roots, social activist groups from which many evolved and that characterize the vast majority of NGOs.[64] Salaries are also distinguishing. The president of the National Wildlife Federation in the US is reported to take home $220,000 per year.[65]

Just as Northern NGOs are becoming more institutionalized, Southern NGOs are building organizational skills and financial independence and, as a result, increasingly demanding greater autonomy and less dependence on Northern supporters.[66] In the multilateral development bank campaign, for example, Southern NGOs have been taking more of the responsibility for setting NGO strategy. In so doing, they have taken up some of the tougher issues of international debt and trade, and even the question of whether the banks should be abolished.[67]

As Southern NGOs are becoming more independent and setting the international agenda, Northern NGOs are looking to the South for ideas, as well as to establish their own international credibility.[68] Thus, although among governments the trend in financial and technology flows has been from the North to the South, the 'technology' of sustainable development is increasingly flowing from South to North.[69] To illustrate, the Grameen Bank, a non-profit, small-loan, self-help development bank in Bangladesh that lends to landless poor and, especially, women, has been so successful (with, for example, 98 per cent repayment rates), that the South Shore Bank of Chicago in the United States is emulating it.[70]

Finally, NGOs vary in the legal status and recognition held in their home countries. In the United States and Europe, citizens take for granted their rights to organize, lobby and protest.[71] In more closed societies, this has not been the case. Nevertheless, as evidenced in Eastern Europe and the former

Soviet Union, environmental activism, especially to the extent NGOs stress public health or economic development concerns, serves as one of the few means of political expression and opposition to the state system.[72] In Indonesia, environmental and development groups are tolerated by an otherwise oppressive regime, as long as they do not challenge state prerogatives. Thus, when some NGOs became too prominent in international lending meetings in Europe, Indonesia clamped down, restricting NGO members' travel rights.[73] In the Philippines in 1991, fourteen members of a leading NGO, the Haribon Foundation, were arrested and charged with subversion.[74] One year later, a priest who was active in anti-illegal logging campaigns was murdered.[75]

In many countries of Africa and, to a lesser extent, Latin America and Asia, NGOs are largely government organized and funded. At UNCED and the parallel Global Forum these groups had to be wary of acting as, or associating with, NGOs which enjoyed greater freedom and were protesting against state policies. In Japan, while NGOs are legal as protest or opposition groups, they are are not widely accepted by society at large. Moreover, most cannot get tax-exempt status because the government's size and budget requirements exceed those of most environmental NGOs.[76]

NGOs' means and goals can often be in conflict, as well. In the 1993 negotiations over the North American Free Trade Agreement, for example, the environmental community in the United States split along fundamental ideological and tactical lines. On one side were the Washington, DC or New York based NGOs such as the National Wildlife Federation (NWF) , the NRDC and the WWF, and, on the other, Sierra Club , FoE, Greenpeace, the American Humane Society and a host of grass-roots organizations. Each side accused the other of distorting facts, failing to recognize economic or political reality, and imperiling public safety. The head of NWF charged that opposing NGOs were 'putting their protectionist polemics ahead of concern for the environment,' while a coalition of opponents asked in an advertisement, 'Why are some "green" groups so quick to sell off the North American environment? Maybe they are too cozy with their corporate funders.'[77]

In sum, international environmental NGOs have a wide range of interests, capabilities, and perspectives. As a result, any attempt to analyse the phenomenon necessarily risks over-generalization. Moreover, the selection of any set of case studies limits the empirical grounding and cannot be adequately representative. Nevertheless, we view this study as a necessary first step, both theoretically and empirically. In Part 1 we draw on existing theoretical literature and deductive reasoning to develop some provisional propositions. In Part 2 we examine in detail a few case histories, rather than survey a smattering of many cases. The objective of both approaches is to begin to build the bases for understanding the status and role of environmental NGOs in world politics, recognizing throughout the diversity of the NGO community.

A necessary first step in this process of building the bases for understanding NGOs is to connect the NGO phenomenon to biophysical changes. Thus, the striking increase in environmental NGOs worldwide and the diversity within the community signal serious trends in ecosystem decline and the concomitant social stress that results from and feeds into that decline. MacNeill, Winsemius, and Yakushiji characterize the crisis as follows:

> The earth's signals are unmistakable. Global warming is a form of feedback from the earth's ecological system to the world's economic system. So is the ozone hole, acid rain in Europe, soil degradation in Africa and Australia, deforestation and species loss in the Amazon. To ignore one system today is to jeopardize the others.[78]

Consequently, a first premise of this book is that a global environmental crisis indeed exists and is worsening.[79] We also assume that despite experience with war, famine, and pestilence, humankind has never before faced environmental problems of this sort, problems that are at once biophysical and social and that have global dimensions. As Young, Demko, and Ramakrishna put it: 'These [environmental] issues present complexities in finding both appropriate conceptualizations of the problems to be solved and satisfactory methods of addressing the equity issues embedded in them. The scope of today's environmental problems is unprecedented.'[80]

Neither traditional diplomacy nor an expanded global economy will be enough to address the challenges of the coming decades. International institutions designed to achieve a balance of power or a stable monetary and trade order are not, *a priori*, capable of responding to the crisis. Moreover, the complex interconnectedness of ecological problems means that institutions with limited sectoral mandates cannot, by themselves, meet the challenges. Environmental problems are no longer simply 'clean-up' problems, nor can they be solved after 'more study' and 'more data.' They involve synergistic and threshold effects, bioaccumulation, multifactorial causation and, most critically, irreversibilities and non-substitutabilities. The uncertainties inherent in such features can easily overwhelm existing institutions.[81]

We thus take urgency and a lack of precedence under conditions of ecological constraint as the primary contextual features of our inquiry. These features suggest that incremental approaches, muddling through, and tinkering around the edges will not be enough. Actors at all levels must change fundamentally how they do business, how they consume products, and how they govern their daily lives and the affairs of states. In short, unlike more familiar crises, where doing more of the same but doing it better may be enough, societies must *learn* their way out of the environmental crisis.[82] And that learning calls for not just education, but a new form of politics, a politics, it turns out, in which NGOs play a role quite different from the accustomed one of lobbying at the national level. NGO emergence is thus indicative of a more profound political transformation.

In this book we therefore assume that change, whether in the form of social learning or political transformation, is necessary in a crisis that is urgent and that has no historical precedent. A focus on change directs attention to the agents of change, agents who envision a sustainable society, demonstrate alternative economic and social systems, and raise public awareness of the trends and their consequences. International environmental NGOs, although not alone in these efforts, appear to be key actors in this regard.[83]

NGOs assume this agent of change role in part because citizens alone or as an unorganized 'movement' cannot. Organization – legal, financial, and political – is essential. To simultaneously reach up to the states and their international institutions and down to the local communities, such agents must establish themselves as independent actors. To provide a counterforce to short-term decision making,[84] to the grow-at-any-price imperative, and to the tendency to find ever more creative means of externalizing costs,[85] they must be able to counter the obvious strengths of governmental and corporate actors without becoming such actors. As argued in Chapter 2, NGOs do this by building bargaining assets as they link the local to the global.

From this perspective, 'education' is not enough, especially if it is carried out by defenders of the status quo. Rather, politically active actors, those who can research, publicize, expose, and monitor environmental trends with little fear of offending constituencies or losing customers, are necessary to promote fundamental change.

The second reason NGOs assume the critical role of agents of social learning is that governments tend not to promote fundamental change, especially change in an economic system dependent on ever-increasing throughput of resources.[86] Indeed, as Young, Demko, and Ramakrishna put it, 'In many social settings, governments are not only poor providers of governance, but they may also be or become instruments of repression, environmental degradation, and bureaucratic paralysis.'[87]

NGOs appear to be key actors in moving societies away from current trends in environmental degradation and toward sustainable economies. Analytically, the choice of NGOs affords a window on the peculiarities of world environmental politics, especially the inadequacies of institutional response. It also offers a sign of hope in the midst of abundant gloom-and-doom scenarios of ecological collapse. That is, we assume that current trends toward ecological collapse are no more inevitable than was nuclear war or the Cold War. Because alternative forms of governance and economic relations are possible, our focus on key agents of change sounds a hopeful note.

At the same time, we approach the subject with caution. Contrary to much popular literature (often promoted by NGOs themselves or sympathetic foundations or even government officials), we do not assume that NGOs are the solution to the global environmental crisis. It is tempting to jump on the bandwagon and declare the rise of NGOs the way of the future.[88] As noted, we share some of the optimistic sentiment that the work of NGOs is hopeful compared to the work of the profit-making and governmental sectors. We do sense that NGOs are at the cutting edge, that, on balance, NGOs are doing good for the planet. When one sees foot-dragging from governments and the corporate sector, excuses for delay and continued destruction, one looks to those who appear to be making a difference.

But NGO relations is a fuzzy area of activity, one quite unlike traditional international diplomacy or domestic policy making. Thus, we cast a critical eye on glowing accounts and facile prescriptions. *A priori*, we do not necessarily accept that NGOs are essential to environmental salvation, nor that they may be anything more than a transitory phenomenon, a wave of activity that helps individuals, organizations, and governments make the transition to new forms of governance consistent with ecological constraints.

We cast this critical eye at two levels. One is the broad picture, the total NGO effort in a given issue or geographic area. Thus, we look at Great Lakes United (GLU) not just as a 'binational' NGO but as a group with basin-wide impact on the entire citizenry of the Great Lakes – Canadians, Americans, and Native Americans. To do that, though, we assess GLU's impact on the respective governments and, particularly, on the international body, the International Joint Commission (IJC). We look at Greenpeace and the Antarctic and Southern Ocean Coalition (ASOC) and examine their ability to upset eight years of careful negotiations by Antarctic Treaty System (ATS) members. We look at the Convention on International Trade in Endangered Species (CITES) and find that this 'international' institution can be profitably viewed as an 'interNGO' institution. In so doing, we do not assume that these groups are 'above politics', that, by virtue of their higher purpose, their 'other-interested' goals, they operate outside of the realm of power and influence.[89] Nor do we assume that they are simply conduits for money and information.[90] Rather, our premise is that they practise a different form of politics, one akin to that of organized religion, some multinational corporations, human rights groups, and others and, yet, somehow different. NGOs do, for example, disseminate important information. But the information is not politically neutral; it is used to promote a political agenda by, among other things, enhancing the transparency of powerful actors.[91] And it is a political agenda grounded in biophysical realities.[92] It is the difference that we attempt to capture in this book.

The second critical level is the micro, the internal management and organization of the NGO. We seek to understand NGO politics in terms of an NGO's organization and the tensions and contradictions it experiences as a non-profit entity. This approach, admittedly, presents special problems. NGOs typically do not open their files for outside inspection. They are, after all, 'private' organizations. For understandable organizational, constituent, and financial reasons, they guard that privacy closely.[93] Nevertheless, the authors of the empirical chapters have managed to penetrate several of the organizations to show how their internal structure determines in part their external politics.

This book, then, is an attempt to gain analytic mileage where traditional approaches and focuses fail. It is an attempt to find more nuanced, more conditional, more contingent prescriptions for environmental policy-making, broadly construed. The NGO window on world environmental politics forces us to ask questions unlikely to emerge from the perspectives of traditional international relations, domestic interest group politics, or social movements. The NGO window also pushes us to ask about results. When a leading student of international environmental policy can claim that 'most problems have already been adequately managed by intergovernmental institutions', we have to wonder what problems analysts are examining and what 'managed' means. In our view, it is not enough to witness treaties signed, conservation plans adopted, and organizations formed. For us, the analytic objective is proportional response, that is, institutional and behavioural responses that correspond to the rates and severity of environmental degradation. Biodiversity loss is not being 'managed' when one hundred species a day are disappearing.

Our concern for results suggests that, to the extent the NGO window helps us focus on change – biophysical, behavioural, and institutional change – a critical analytic task is to understand the relevant change processes. To do this we need both conceptual and empirical development.

In Part 1 of the book we critically assess two theoretical perspectives – political bargaining and social movements – for their ability to characterize the origins, growth, and politics of NGOs. Both are decidedly non-state perspectives on world politics.[94] But, because they do not fully capture the NGO phenomenon, we develop new concepts to account for it.

In Part 2 we supply the second ingredient for understanding change processes – empirical grounding. Abstract notions of NGO relations must be built on and tested against the events as they occur. Were this a study of, say, crisis bargaining or trade protectionism, we could simply pull from any library shelf volumes of detailed case histories with which to ground our theorizing. Moreover, if we even mention the Cuban missile crisis or nineteenth-century mercantilism, readers would share common empirical referent points. Such is not the case in the realm of world environmental politics. Diplomatic histories are meagre in part because the field itself – in study and in practice – is so new. Environmental historians are just now beginning to move beyond domestic or comparative histories to international histories.[95] Moreover, perusing any bookshop, one will see shelves full of books on the state of the environment and on what people should do to save the planet. Precious little exists, however, on the tough choices decision makers – governmental and non-governmental – must make in trying to protect environmental values, seek new forms of governance, and foster social and environmental learning.

The case histories were chosen for this book, therefore, because they are rich in NGO relations and diverse in their subject matter and geographic scope. They do not, of course, constitute a representative sample of NGO activities but they do serve several functions. They illustrate much of the NGO phenomenon and, in particular, the interactions among NGOs and governments and intergovernmental bodies. They are examples of 'thick description', accounts that give enough detail for the reader to appreciate the context and the challenges decision makers face when resolving complex environmental problems. Consequently, they are lengthy. Second, in these case studies we attempt to reveal the nuance, the conditionality, the ambiguity, and the uncertainty of NGO relations. So, wherever possible, the case writers have endeavoured to give the full flavour of the situation with, for example, extensive description of the social, political, and biophysical context and generous quoting of key players.

A third and, arguably, most important, function of the case studies is to generate new questions, questions about NGO relations, about world environmental policy-making, and about world politics generally. Each author uses the case study and parts of the theoretical approaches in Part 1, to explore the implications, the extensions, and the general lessons of the case. Because this is a book designed to stimulate new thinking, the authors at times take the implications to the limit. These implications are not intended as definitive statements. Rather, our assumption is that the state of knowledge regarding world environmental politics is in such a stage of infancy that what is needed now is wide-ranging ideas and propositions, even if most are ultimately shown to be 'wrong'. It is not possible, for example, to take one of the theories or one set of propositions of Part I and test them against the cases in Part 2. We return to many of these implications and propositions in the conclusion to show where they may lead and how they might be provisionally integrated. The comparative analysis then becomes clearer.

The cases were chosen in part because of the richness of NGO activity and in part because they span wide geographic and substantive areas. They are all instances of transboundary environmental problems and transnational politics including interactions with international organizations. They cannot pretend to be, however, representative of the entire NGO phenomenon. By choosing a few cases to develop in depth we have necessarily limited the scope of NGO relations. We nevertheless are confident that in an emerging field of study with a thin empirical base, this is a necessary and useful start.

If a 'theory' emerges in this study, it is only to the extent that we have assembled a loose set of propositions and concepts with preliminary empirical grounding and testing. We fully believe that a tight, logical theory of world environmental politics, one grounded in the biophysical and social realities of the 1990s where ecological constraint, urgency, and a lack of precedent are primary contextual features, is still a long way away. We hope, however, to contribute to the development of such a theory by critically examining the work of NGOs, one set of actors who many observers see as critical agents of progressive change.

NOTES

1 To illustrate, Oran R. Young, in his seminal work in the field of international environmental politics, *International Cooperation: Building Regimes for Natural Resources and the Environment* (Ithaca, NY: Cornell University Press, 1989), devotes less than two pages to NGOs, noting how regimes give rise to such groups and how NGOs defend the provisions of regimes (78). But in conference proceedings several years later, he and two colleagues repeatedly cite the critical role of NGOs, saying that they

> now loom large not only in processes of regime formation but also in catalyzing and aggregating public pressure on officials to live up to the commitments they make. The environmental movement, once concentrated almost exclusively on domestic concerns, has become a force to be reckoned with in the political dynamics surrounding international environmental governing.
>
> (Oran R. Young, George J. Demko, and Kilaparti Ramakrishna,
> 'Global Environmental Change and International Governance'
> (Summary and recommendations of a conference held at
> Dartmouth College, Hanover, NH, June 1991) 13

Lynton K. Caldwell was probably the first to call scholars' attention to the NGO phenomenon and to document its rise. He updates and interprets this phenomenon in a 1988 book chapter in which he states that NGOs have been 'absolutely essential to most international environmental action' and the 'nature and extent of NGO influence on international environmental policy has not received comprehensive or detailed study': 'Beyond environmental diplomacy: the changing institutional structure of international cooperation', in John E. Carroll, ed., *International Environmental Diplomacy* (Cambridge: Cambridge University Press, 1988) 24.

In one of the first attempts to explicitly link global environmental issues to questions of security and political economy, Caroline Thomas goes as far as saying 'that the current diplomatic profile of environmental issues derives largely from the activities of NGOs who took advantage of the political space provided by the fortuitous ending of the Cold War', *The Environment in International Relations* (London: The Royal Institute of International Affairs, 1992) 14.

2 We use 'world environmental politics' to denote both *transboundary* and *transnational* environmental politics. Thus, the term includes but is not limited to global politics and, unlike the term 'international', it does not connote the more restrictive usage of interstate and intergovernmental politics. In using *transnational* we borrow from Thomas Risse-Kappen's definition of transnational relations in *Bringing Transnational Relations Back In* (forthcoming): 'regular interactions across national boundaries when at least one actor is a non-state agent or does not operate on behalf of a national government or an intergovernmental organization.' These interactions include transgovernmental (subunits of state governments) and transsocietal (ethnic or other distinct social groups that live within or across state boundaries) relations.

3 Reliable figures for the numbers of NGOs are difficult to procure. Records have not been regularly kept except for officially recognized NGOs. Another reason is that no one can agree on what exactly an NGO is. 'Non-governmental' organizations span a continuum from those that are purely voluntary groups with no governmental affiliation or support, to those that are creations and arms of governments. Moreover, many are highly market driven or are creations of corporations so that they are, for all practical purposes, *profit-making* 'non-governmental' groups. Finally, many 'environmental' groups are as much

development or human rights or social justice groups, especially in the less-industrialized countries. For further discussion, especially from an international development perspective, see David Korten, *Getting to the 21st Century: Voluntary Action and the Global Agenda* (West Hartford, CT: Kumarian Press, 1990) 2–3, 94–105.

In this book, we use the term 'environmental NGO' to denote those non-profit groups whose primary mission is to reverse environmental degradation or promote sustainable forms of development, not to pursue the objectives of governmental or corporate actors. 'International environmental NGOs' refers to those groups with bases or activities in more than one country. Because international environmental NGOs are the topic of this study, we use 'NGO' as a shorthand designation, unless clarification is deemed necessary.

4 Union of International Associations, *Yearbook of International Organizations 1988/89* (Munich: Saur Verlag, 1988) 36.

5 World Resources Institute in collaboration with the United Nations Environment Programme and the United Nations Development Programme, *World Resources 1992–93: A Guide to the Global Environment* (New York: Oxford University Press, 1992) 216.

6 The Economic and Social Council at the UN accredits NGOs directly to the UN and to some of its specialized agencies. However, of the some 460-odd accredited NGOs, which include such groups as the World Conservation Union, the Sierra Club, the National Audubon Society, and the World Resources Institute, less than 5 per cent are environmental (personal communication by Finger, United Nations NGO Liaison Office, Geneva, 1993).

7 Mostafa Tolba, Osama El-Kholy, E. El-Hinnawi, M.W. Holdgate, D.F. McMichael, and R.E. Munn, eds, *The World Environment 1972–1992: Two Decades of Challenge* (London: Chapman and Hall, 1992) 681.

8 Tolba *et al.*, The World Environment *1972–1992*, 725.

9 Tolba *et al.*, The World Environment *1972–1992*, 728.

10 *Earth Summit News*, May 26, 1993, Econet, electronic news, topic 744, San Francisco, CA: Institute for Global Communications.

11 Tolba *et al.*, *The World Environment 1972–1992*, 729.

12 Tolba *et al.*, *The World Environment 1972–1992*, 725.

13 James V. Riker, 'Linking Development from Below to the International Environmental Movement: Sustainable Development and State–NGO Relations in Indonesia', paper presented at annual meeting of Northwest Regional Consortium for Southeast Asian Studies on 'Development, Environment, Community and the Role of the State', 16–18 October 1992, University of British Columbia, Vancouver, Canada) 12, 14.

14 World Resources Institute, *World Resources 1992–93*, 218.

15 World Resources Institute, *World Resources 1992–93*, 218.

16 Philip R. Pryde, *Environmental Management in the Soviet Union* (Cambridge: Cambridge University Press, 1991) 253.

17 Thaddeus C. Trzyna and Roberta Childers, eds. *World Directory of Environmental Organizations*, 4th edn (Sacramento, CA: California Institute of Public Affairs, 1992), in cooperation with the Sierra Club and The World Conservation Union (IUCN); Worldwise and Friends of the Earth, *International Directory of Non-Governmental Organizations* (Sacramento, CA: Worldwise and Friends of the Earth, 1992). Other examples of recent directories include Todd Nachowitz, ed., *An Alternative Directory of Nongovernmental Organizations in South Asia*, rev. edn. (Syracuse, NY: Maxwell School of Citizenship and Public Affairs, Syracuse University, 1990; distributed by Foreign and Comparative Studies South Asian Studies and Fourth World Press).

18 H. Keller, ed., *Who is Who in Service to the Earth*, (Waynesville, NC: VisionLink Foundation, 1991).

19 We use 'Northern' to denote industrialized countries from North America and Western Europe plus Japan, Australia and New Zealand. 'Southern' refers to those less industrialized countries formerly known as the Third World and occupying much of Latin America, Africa, and Asia.

20 With world-wide economic recession in the early 1990s, many of these organizations have experienced tighter budgets, and lay-offs.

21 World Wildlife Fund-US (annual report, 1991).

22 Joy Meeker, 'Greenpeace' (paper, Syracuse University, NY, 1991); Bob Ostertag, 'Greenpeace Takes Over the World', *Mother Jones* (March/April 1991): 85.

23 Tolba *et al.*, *The World Environment 1972–1992*, 680.

24 *Friends of the Earth*, newsletter of Friends of the Earth-US, Washington, DC, 23 (1) (January 1993): 14; Chris Sneddon and Alma Lowry, 'Friends of the Earth International: The Role of an International Network in Global Environmental Politics' (paper, April 1992, School of Natural Resources and Environment, University of Michigan).

25 Frederick J. Long, 'The Nature Conservancy', case number S-PM-32 (Stanford, CA: Graduate School of Business, Stanford University, 1992) 2.

26 R. Mitchell *et al.*, 'Twenty Years of Environmental Mobilization: Trends among National Environmental Organizations.' *Society and Natural Resources*, vol. 4 (1991) 219–34.

27 For further characterization of United States NGOs, their growth and their increasing internationalization, see Barbara J. Bramble and Gareth Porter, 'Non-Governmental Organizations and the Making of US International Environmental Policy', in Andrew Hurrell and Benedict Kingsbury, eds, *The International Politics of the Environment: Actors, Interests, and Institutions* (Oxford: Oxford University Press, 1992) 313–53.

28 David C. Korten, *Getting to the 21st Century: Voluntary Action and the Global Agenda* (West Hartford, CT: Kumarian Press, 1990) 93, 94.

29 Riker, 'Linking Development from Below to the International Environmental Movement' (21–2).

30 *Japanese Working for a Better World: Grassroots Voices and Access Guide to Citizens' Groups in Japan* (San Francisco, CA: HONNOKI USA, 1992) 5.

31 ENDA Tiers-Monde, 'Emergency Fight on Poverty, for Democracy and the Environment: The New Frontier', typescript, Dakar, Senegal (undated, but distributed at the 1992 Global Forum in Rio de Janeiro, Brazil).

32 Gail Osherenko and Oran R. Young, *The Age of the Arctic: Hot Conflicts and Cold Realities* (Cambridge: Cambridge University Press, 1989) 91–109; see also Margaret L. Clark and John S. Dryzek, 'The Inuit Circumpolar Conference as an International Nongovernmental Actor', paper prepared for the Arctic Policy Conference at McGill University, Canada, September 19–21, 1985.

33 Alan Thein Durning, 'Native Americans Stand Their Ground,' *Worldwatch*, November/December, 1991, 10–17.

34 Jagdish Parikh, 'Indigenous-Tribal Forest Peoples' Alliance Aims to Halt "Ecocide and Ethnocide", March 26, 1992, by NGONET on Econet conference, cdp:en.unced.gener.

35 For discussion of existing and potential roles of NGOs in the Middle East, see Munir Adgham, 'Non-Governmental Environmental Organizations in the Gulf of Aqaba-Bordering States: A Current Appraisal' (479–89); Omnia Amer, 'An Agenda for Cooperation Among Non-Governmental Organizations in the Gulf of Aqaba-Bordering States' (491–7); and Thomas Princen, 'The Role of Non-Governmental Organizations in Fostering International Environmental Governance' (463–78) all in Environmental Law Institute, *Protecting the Gulf of Aqaba:*

A Regional Environmental Challenge (Washington, DC: Environmental Law Institute, 1993).

On Israeli environmentalism, see Penina M. Glazer and Myron P. Glazer, 'Citizens' Crusade for a Safe Environment: Israel's Determined Few', paper presented at the eighth annual Israel Conference Day, January 31, 1993, University of Michigan, Ann Arbor, MI. This is part of a larger project comparing environmentalism in Israel, the United States, the Czech Republic and Slovakia.

36 'The European Environmental Bureau: European Federation of Environmental NGOs', eight-page summary statement, no date (distributed 1992), Brussels, Belgium.

37 'East/West Non-Governmental Organizations Action Plan for the Ecological Reconstruction of Central and Eastern Europe', statement of the Vienna Conference on the Ecological Reconstruction of Central and Eastern Europe, November 15–17, 1992; organizer: Global 2000, co-organizers: Greenway, Friends of the Earth European Coordination.

38 'US–Mexico Border Ecojustice Network Meets', *Citizens Network for Sustainable Development Newsletter*, New York: Number 20, September 1993 (1).

39 *Interaction*, newsletter of Global Tomorrow Coalition, Washington, DC, 10 (3) (Summer 1991): 4.

40 Gareth Porter and Janet Welsh Brown, *Global Environmental Politics* (Boulder, CO: Westview, 1991) 59.

41 US Citizens Network newsletter, July 2, 1992.

42 'Reporters' Notebook', *Earth Summit Times*, official newspaper of record of UNCED (New York/Rio de Janeiro, June 5, 1992) 16. The figure could not be substantiated elsewhere and is probably exaggerated, since the term NGO in United Nations parlance means any group that is literally non-governmental, including business and trade unions.

43 Charlotte Patton, 'NGO news from the United Nations', ISA International Studies Newsletter, September 1993, 20 (7)1, 15.

44 'NGOs Unite to Stop Overfishing', *Citizens Network for Sustainable Development Newsletter*, New York: Number 20, September 1993 (1,4). Among the NGOs were Greenpeace, WWF, NRDC, EDF, National Audubon society, American Oceans Campaign, and several Canadian NGOs. The network published daily *Earth Negotiations Bulletin* and six issues of *Eco*, a widely circulated newsletter.

45 Charlotte Patton, 'NGO news from the United Nations', ISA International Studies Newsletter, September 1993, 20 (7)1.

46 Nancy Lindborg is less generous in her assessment of parallel conferences, characterizing them as 'jamborees' and 'theatrical exercises' where NGOs influence delegates only indirectly through the media and through demonstrating their ability to mobilize a large constituency. 'Nongovernmental Organizations: Their Past, Present, and Future Role in International Environmental Negotiations', in Lawrence E. Susskind, Eric Jay Dolin and J. William Breslin, *International Environmental Treaty Making* (Cambridge, MA: Program on Negotiation at the Harvard Law School, 1992).

We, too, take a cautious view of the utility of parallel conferences. At the same time, we attribute significance to these exercises for their contribution to overall social learning (not just to the effects on delegates, or to the self-education of NGO representatives) and to the quasi-institutionalization of the global NGO community as it links the local and the global. See Chapters 7 and 8.

47 Regarding the evolution of the ITTO's conservation goals, see Thomas Princen, 'From Timber to Forest: The Evolution of the Tropical Timber Trade Regime', typescript, University of Michigan, Ann Arbor, MI. For a short case description of the ITTO and the role of NGOs, see Barbara J. Bramble and Gareth Porter, 'Non-Governmental Organizations and the Making of US International Environmental

Policy', in Andrew Hurrell and Benedict Kingsbury, eds, *The International Politics of the Environment: Actors, Interests, and Institutions* (Oxford: Oxford University Press, 1992) 313–53.

48 'ITTA renegotiation sees little progress', *ISA Newsletter*, Provo, UT: International Studies Association, 20 (6) August 1993, 13–14.

49 Patricia Birnie, 'The Role of International Law in Solving Certain Environmental Conflicts', in John E. Carroll, ed., *International Environmental Diplomacy* (Cambridge: Cambridge University Press, 1988) 108.

Whaling states have since adopted the same strategy, building up their side such that in 1993 they could at least block major actions that require a three-quarters majority in the IWC. Andrew Pollack, 'Agency Takes Step Toward Banning Whaling in Southern Ocean' (*New York Times*, May 15, 1993) 2.

50 Kevin Stairs and Peter Taylor, 'Non-Governmental Organizations and Legal Protection of the Oceans: A Case Study' in Andrew Hurrell and Benedict Kingsbury, eds, *The International Politics of the Environment: Actors, Interests, and Institutions* (Oxford: Oxford University Press, 1992) 110–41.

51 A similar effort occurred on the Continent, with the European Environment Bureau and the European Bureau of Consumers' Unions taking the lead. The ultimate result was more encompassing, however, as the pressure on industry and on governments led to adoption without changes by the European Community of the Montreal protocol as binding legislation. For details of the critical role NGOs, scientists and the media played in this process, see M. Leann Brown, 'Agenda Setting, Policy Making, and Institutional Learning in an International Setting: The Greening of the European Community', paper presented at 1992 annual meeting of the International Studies Association, Atlanta, GA. For details of NGO involvement in the entire ozone protection effort, see Barbara J. Bramble and Gareth Porter, 'Non-Governmental Organizations and the Making of US International Environmental Policy', in Andrew Hurrell and Benedict Kingsbury, eds, *The International Politics of the Environment: Actors, Interests, and Institutions* (Oxford: Oxford University Press, 1992) 313–53; Elizabeth Cook, 'Global Environmental Advocacy: Citizen Activism in Protecting the Ozone Layer', *Ambio* 19, (6–7) (October 1990) 334–7; and Richard E. Benedick, *Ozone Diplomacy: New Directions in Safeguarding the Planet* (Cambridge, MA: Harvard University Press, 1991).

For a more critical perspective that attributes the movement on the ozone agreement to media sensationalism and NGO opportunism, see Allan Mazur and Jinling Lee, 'Sounding the Global Alarm: Environmental Issues in the National News', *Social Studies of Science*, 23 (1993) 681–720.

52 Interview by Princen of officials of Legal Rights and Natural Resources Center – Kasama sa Kalikasan/Friends of the Earth, Philippines, Quezon City, Philippines, 1992.

53 World Resources Institute, *World Resources 1992–93*, 217.

54 For example, a January 1993 meeting on biodiversity included the World Resources Institute, the World Conservation Union, the World Wildlife Fund, Conservation International and the Nature Conservancy. *Environment Bulletin*, newsletter of the World Bank environment department, Washington, DC, 5 (2) (Spring 1993):8. With respect to the GEF, Bank policy is to include NGOs throughout the development and implementation of the Facility. In 1992, a steering committee composed of NGOs and implementing agencies was formed. *Environment Bulletin*, 4 (4) (Fall 1992):5. For discussion of the developments leading to and including such NGO participation, see Pat Aufderheide and Bruce Rich, 'Environmental Reform and the Multilateral Banks', *World Policy Journal*, 5 (2) (1988); and Barbara J. Bramble and Gareth Porter, 'Non-Governmental Organizations and the Making of US International Environmental Policy', in

Andrew Hurrell and Benedict Kingsbury, eds, *The International Politics of the Environment: Actors, Interests, and Institutions* (Oxford: Oxford University Press, 1992) 313–53. For an assessment of a useful relationship between development NGOs and the Bank, see L. David Brown and David C. Korten, 'Understanding Voluntary Organizations: Guidelines for Donors', WPS 258, Working Paper, Country Economics Department, The World Bank, Washington, DC (September, 1989).

For a critical assessment of the GEF and NGO participation, see Oliver Tickell and Nicholas Hildyard, 'Green Dollars, Green Menace', *The Ecologist*, 22 (3) (May/June 1992): 82–3.

It should be added that although many NGOs continue to invest organizational resources in monitoring and working with such international organizations as the ITTO and the World Bank, many others, including those with considerable experience, have forsaken the possibility of reforming these organizations, let alone of their being sources of necessary change (personal communication by Princen and members of Japanese NGOs with respect to the ITTO and David C. Korten with respect to the World Bank, 1992).

55 'FAO Special Issue', *The Ecologist*, 21 (2) (March/April 1991).

56 World Resources Institute, *World Resources 1992–93*, 224.

57 *Japanese Working for a Better World: Grassroots Voices and Access Guide to Citizens' Groups in Japan* (San Francisco, CA: HONNOKI USA, 1992) 5.

58 The relative stability of large international NGOs may still belie their seemingly constant state of flux, especially with respect to financing and membership. In fact, at the organizational level, a defining characteristic of many big NGOs may be that they are constantly defining themselves. Unlike most governmental and, depending on markets, business organizations, there is no resting point. NGOs must constantly adjust to the whims of the public, political leaders and donors, not to mention adjusting to challenges presented by ever more extensive and complex environmental problems.

59 Sergio Guillen, 'The Role of US-Mexico Border Communities as Actors in Transboundary Environmental Policy', paper, University of Michigan, 1991.

60 For a lucid discussion of the more radical ideologies, see Carolyn Merchant, *Radical Ecology: The Search for a Livable World* (London: Routledge, 1992).

61 The extensive organizing nationally among Filipino NGOs was extended to the 1992 Global Forum in Rio, where Filipinos were prominent leaders organizing NGOs globally.

62 A recent survey of 1,000 groups in Brazil found that 47 per cent were independent, 37 per cent had church affiliations, and the remainder were linked to institutions such as universities. Also, 35 per cent of the organizations were started in the 1970s and 55 per cent in the 1980s. World Resources Institute, *World Resources 1992–93*, 14.

63 Although most women's NGOs are small, grass-roots organizations, some are prominent on the international scene, in particular, the Green Belt Movement in Kenya, led by Wangari Maathai and CEFEMINA, Feminist Centre for Information and Action of Costa Rica. See Paul Ekins, *A New World Order: Grassroots Movements for Global Change* (London: Routledge, 1992) on CEFEMINA, pp. 78–80, and on Green Belt, pp. 151–2.

64 Although the great majority of Japanese NGOs are still small, grass-roots groups organized to oppose specific development projects or industries within Japan, some are internationalizing and developing corresponding organizational cultures. For example, WWF-Japan and Friends of the Earth Japan, established in 1971 and 1979 respectively, build working relationships with business and government and are themselves organized more hierarchically than grass-roots groups. W. Puck Brecher, 'CMs [Citizen Movements] and NGOs: Grassroots Perspectives on the

Japanese Environmental Gridlock', masters thesis, Asian Studies, University of Michigan, August 31, 1992.

65 John Lancaster, 'Jay Hair's Environmental Impact: Playing Hardball at the National Wildlife Federation', *The Washington Post National Weekly Edition*, September 9–15, 1991: 12–13.

66 By 1989, Northern NGOs were distributing an estimated $6.4 billion to the South, about 12 per cent of all development aid, public and private. This means that Northern NGOs collectively distribute more than the World Bank. World Resources Institute, *World Resources 1992–93*, 218.

67 Barbara J. Bramble and Gareth Porter, 'Non-Governmental Organizations and the Making of US International Environmental Policy', in Andrew Hurrell and Benedict Kingsbury, eds, *The International Politics of the Environment: Actors, Interests, and Institutions* (Oxford: Oxford University Press, 1992) 349.

68 In interviews with numerous NGO leaders in the Philippines, Princen observed that such groups are besieged with requests from northern NGOs, researchers, foundation representatives and the like. Many of these visitors seek information on organizing techniques and conceptual developments in the area of sustainable development. In this respect, the flow of 'technology' is the reverse of that of traditional North–South flows.

69 Drawing on evidence of traditional environmental regulations and procedures in African societies, David R. Penna argues that knowledge transfers need to be two-way, the North providing scientific training and the South non-economic incentive schemes. 'Regulation ofthe Environment in Traditional Society as a Basis for the Right to a Satisfactory Environment', paper presented at the annual meeting of the International Studies Association, Mexico, 1993.

70 Ekins, *A New World Order*, 122–7.

71 Despite the relatively more open societies in North America and Europe, in the United Kingdom, for example, NGOs were unable to influence pollution policies, owing to a pattern of secretly negotiated agreements between government and industry with compliance only voluntary. With admission to the European Community and its mandatory standard-setting policies, a channel opened for public participation. See John McCormick, 'British Environmental Policy and the European Community', paper presented at 1991 annual meeting of the International Studies Association, Vancouver, BC, Canada.

Linkages between national, even local, NGOs and intergovernmental organizations are a common pattern in NGO relations and are discussed in depth in the Great Lakes case (Chapter 4) and the conclusion (Chapter 8).

72 See, for example, Barbara Jancar-Webster, 'Chaos as an Explanation of the Role of Environmental Groups in East Europe Politics' (paper presented at the annual meeting of the International Studies Association, London, March 1989).

73 Personal communication, Ann Hawkins, 1992. Riker argues that

the environmental arena has provided NGOs with a safe and acceptable basis by which to organize. By choosing a sectoral area where international NGOs enjoy considerable legitimacy and where the Government of Indonesia lacks expertise, especially at the field level, Indonesian NGOs have gained a foothold in the political sphere.

('Linking Development from Below to the International Environmental Movement') 6.

74 'Philippine Environmentalists Arrested for Efforts to Save Rainforest', *Haribon Update*, newsletter of the Haribon Foundation, 6 (2) (March–April 1991) 1, 4.

75 'Anti-logging Priest Killed in Ambush', *Haribon Update*, newsletter of the Haribon Foundation, 7 (1) (January–February 1992) 2,3, 4–5.

76 This observation results from interviews Princen made, mostly confidential, with Japanese environmentalists, scholars and journalists in Tokyo, Japan, 1992. It is also noteworthy that when the Japanese government organized the International Conference on Global Environmental Protection Towards Sustainable Development in Tokyo in 1989, NGOs were prevented from participating. Brecher, 'CMs [Citizen Movements] and NGOs', 1992.

77 Keith Schneider, 'Environmentalists Fight Each Other Over Trade Accord', *New York Times*, September 16, 1993, A1, A10.

78 Jim MacNeill, Pieter Winsemius, and Taizo Yakushiji, *Beyond Interdependence: The Meshing of the World's Economy and the Earth's Ecology* (New York: Oxford University Press, 1991) 4.

79 The literature documenting these trends is voluminous. See, for example, Lester R. Brown *et al.*, *State of the World* (New York: Norton, 1992); B.L. Turner II, William C. Clark, Robert W. Kates, John F. Richards, Jessica T. Mathews, and William B. Meyer, eds, *The Earth as Transformed by Human Action: Global and Regional Changes in the Biosphere over the Past 300 Years* (Cambridge: Cambridge University Press, 1990, with Clark University); World Resources Institute, *World Resources 1992–93*, (New York: Oxford University Press, 1992).

80 Young, Demko, and Ramakrishna, 'Global Environmental Change', 7.

81 For a discussion of the limitations of 'assimilative capacity' arguments and the shift toward precautionary approaches – a shift largely promoted by environmental NGOs – see Kevin Stairs and Peter Taylor, 'Non-Governmental Organizations and Legal Protection of the Oceans: A Case Study' in Andrew Hurrell and Benedict Kingsbury, eds, *The International Politics of the Environment: Actors, Interests, and Institutions* (Oxford: Oxford University Press, 1992) 110–41. For analysis of the precautionary principle from a decision analytic perspective where traditional science is seen as inadequate and ethical judgements ubiquitous, see Charles Perrings, 'Reserved Rationality and the Precautionary Principle: Technological Change, Time and Uncertainty in Environmental Decision Making', in Robert Costanza, ed., *Ecological Economics: The Science and Management of Sustainability* (New York: Columbia University Press, 1991) 153–66. Perrings concludes (p.164):

> As our knowledge of the global system increases, so does our uncertainty about the long term implications of present economic activity. Combined with the uncertainty caused by the rapid pace of change in resource use technology, this suggests that the increasing flow of information does not in fact give more complete information. The problem for decision makers does not get easier. Not only is the perceived range and severity of the possible environmental effects of economic activity expanding, so is the gestation period.

82 See Chapter 3 for elaboration of the concept of social learning and its connection to environmental crisis.

83 Among key actors which perform similar catalytic functions are private funding agencies such as the Ford Foundation (see Korten, *Getting to the 21st Century*); quasi-governmental agencies such as the Smithsonian Institution in the United States (Leonard P. Hirsch, presentation at the 1993 annual meeting of the International Studies Association, Mexico); and intergovernmental agencies such as the United Nations Environment Programme (for detailed accounts in two issue areas, see Benedick, *Ozone Diplomacy*; and Peter M. Haas, *Saving the Mediterranean: The Politics of International Environmental Cooperation*, New York: Columbia University Press, 1990) and the World Meteorological Organization (see Marvin S. Soroos and Elena N. Nikitina, 'The World Meteorological Organization as a Purveyor of Global Public Goods', paper presented at the annual meeting of the International Studies Association, Mexico, 1993).

84 Garry D. Brewer, 'Environmental Challenges and Managerial Responses', in Nazli Choucri, ed., *Global Accord: Environmental Challenges and International Responses* (Cambridge, MA: MIT Press, 1993) 281–305.

85 Thomas Princen, 'Ivory, Conservation and Transnational Environmental Coalitions', in Thomas Risse-Kappen, ed., *Bringing Transnational Relations Back In*.

86 The emerging field of ecological economics challenges the assumptions and methodologies of conventional, neoclassical economics. A core premise in the field is that the economic system is a subset of a finite biophysical system where the throughput of resources and energy cannot expand forever. See Robert Costanza, ed., *Ecological Economics: The Science and Management of Sustainability* (New York: Columbia University Press, 1991); Herman Daly, *Steady-State Economics* (Covela, CA: Island Press, 1991).

87 Young, Demko, and Ramakrishna, 'Global Environmental Change', 6. Caldwell elaborates on the inadequacies of national governments and their dilemma involved in achieving international environmental cooperation:

> National governments are characteristically reluctant, and seldom able on their own motion, to initiate proposals for international environmental cooperation. Yet effective international cooperation for environmental protection requires action or abstinence by national governments. Negative inclinations toward positive commitments are not unique to environmental issues, but are strongly characteristic of them. Such issues tend to extend geographically beyond any one nation's frontiers and beyond the lifetime of political office-holders and their constituents. National leaders are therefore unlikely to see much advantage in the serious pursuit of international environmental cooperation unless their own national (and personal) interests are involved.
>
> (Carroll, ed., *International Environmental Diplomacy*, 1988) 14.

88 One of the few areas of NGO relations that receive any critical analysis is international conservation. In much of this work, authors see intimate and, ultimately, destructive linkages between NGOs and the state, especially the state's military and other security forces. See Chapters 5 and 7 and, for example, David Barkin and Steve Mumme, 'Environmentalists Abroad: Ethical and Policy Implications of Environmental Non-Governmental Organizations in the Third World', paper prepared for the Third Congress of the International Development Ethics Association, Tegucigalpa, Honduras, 21–8 June 1992; Susanna Hecht and Alexander Cockburn, *The Fate of the Forest: Developers, Destroyers and Defenders of the Amazon* (New York: Harper Perennial, 1990); Hira Jhamtani, 'The Imperialism of Northern NGOs', *Earth Island Journal* (San Francisco, CA: Earth Island Institute) vol. 7 (June 1992) 10; Matthias Finger, 'The Military, the Nation State and the Environment', *The Ecologist*, 21 (5), (September/October 1991) 220–5; Nancy Lee Peluso, 'Coercing Conservation: The Politics of State Resource Control', paper presented at 1992 annual meeting of the International Studies Association, Atlanta, GA.

89 NGOs themselves commonly perpetuate an apolitical view of NGO relations – and for good *political* reasons. In a book written largely by conservation and development NGOs, Alison Jolly's assessment is characteristic:

> The enormous importance of NGOs lies in their expertise. It is they who know scientists, both foreign and local. It is their people who have slept in the woods, birdwatched at dawn, negotiated at noon with the provincial politician, and exchanged oratorical speeches of friendship with the local chief under the moon, washed down in local beer. This style may not cut much ice in Washington, but people who link diverse worlds, and who know and work with

the local conservation professionals as friends and colleagues, not clients, are the ones who will in the end be respected.

'The Madagascar Challenge: Human Needs and Fragile Ecosystems', 189–215, in H. Jeffrey Leonard, ed., *Environmental and the Poor: Development Strategies for a Common Agenda*, New Brunswick, NJ: Transaction Books, 214.

Much of the official rationale by governments for NGO involvement in UNCED and other intergovernmental meetings is that NGOs are important disseminators of governmental information and funds. Ann Hawkins sees this tendency as part of a global managerialist paradigm where a division of labour between governments and international organizations places NGOs 'in a kind of advisory role on the sidelines' ('Contested Ground: International Environmentalism and Global Climate Change', paper presented at 1992 annual meeting of the International Studies Association, Atlanta, GA.) On the global management perspective and the UNCED process, see Chapter 7.

90 For cognitive approaches to the role of NGOs, approaches that do account for the political impact of non-state actors, see the work on 'epistemic communities', especially Peter M. Haas, ed., *Knowledge, Power, and International Policy Coordination* (special edition), *International Organization* 46 (1) (Winter 1992); and, from a Southern perspective, Shimwaayi Muntemba, 'Research for Sustainable Development: The Role of NGOs', *Development* 2 (3) (1989) 65–7. In an argument consistent with the theme of this book, Ronnie D. Lipschutz argues that the epistemic community approach still has a state-centric focus and does not account for the fact that such action can operate in non-governmental contexts. This is especially true in oppositional and parallel transnational activities such as the Natural Resources Defense Council's seismic monitoring project in the former Soviet Union and the many debt-for-nature swaps. 'From Here to Eternity: Environmental Time Frames and National Decisionmaking', paper presented at the 1991 annual meeting of the International Studies Association, Vancouver, BC, Canada, 1991.

91 On the political role of enhancing transparency, see Chapter 2 and Karen Mingst, 'Implementing International Environmental Treaties: The Role of NGOs', paper presented at the annual meeting of the International Studies Association, Mexico, 1993.

92 What makes environmental NGOs different from other NGOs – eg., human rights, health, feminist, gay rights – is this connection to the biophysical. For elaboration of this point, see Chapter 8.

93 The difficulty of gaining access to the inner workings of NGOs is a problem not only for scholars but also for grass-roots groups, which regularly decry the lack of transparency and openness of the big, capital-city NGOs. For discussion of this issue in the United States, see Riley Dunlap and Angela Mertig, *American Environmentalism: The US Environmental Movement, 1970–1990* (Washington, DC: Taylor and Francis, 1992).

94 The methodological justification for beginning with non-state-centric approaches is that to understand the peculiarities of environmental affairs at the world politics level, it is necessary to seek analytic starting points different from those traditionally applied to issues of military security and trade and monetary relations. Our premise is that a hierarchy of issue areas in international affairs does not necessarily exist – certainly not, as is customary, a hierarchy that extends from military security to economic production to human rights, democracy, and the environment. Each area warrants its own independent analytic attention, and each is inextricably linked with the others. Probably no single issue area can match environment for its analytical and policy-oriented need to be inclusive, eclectic,

integrative, and multidisciplinary. To start with concepts developed for security questions, such as balance of power or alliance politics, or for economic relations, such as the gains from trade or the global market-place, only skews the analysis and, hence, the prescriptions toward variations on traditional conflict management and world order notions. Thus, we endeavour in this book to break out of those modes while we build on established intellectual traditions that do not necessarily channel our thinking along traditional, state-centric, issue-hierarchical lines.

95 See, for example, John F. Richards and Richard Tucker, eds, *World Deforestation in the Twentieth Century* (Durham, NC: Duke University Press, 1988); Donald Worster, ed., *The Ends of the Earth: Perspectives on Modern Environmental History* (Cambridge: Cambridge University Press, 1988).

Part I
Theoretical perspectives

2 NGOs: creating a niche in environmental diplomacy

Thomas Princen

Of the many approaches to analysing questions of global peace and prosperity, two dominant approaches can be discerned. One can be termed 'top-down' and corresponds to traditional forms of diplomacy with roots in the European classical balance of power security systems.[1] The emphasis is on states, especially powerful, industrialized states, and their financial and trade institutions. The other can be termed 'bottom-up' and emphasizes grass-roots organizing, participatory decision making, and local self-reliance. Together, the top-down and bottom-up approaches capture a range of international activities related to international political economy and international development. But when applied to the environmental realm, they tend to miss key ingredients in the policy-making process – namely, links between the local and the international, on the one hand, and evolving governance structures characterized by anticipation, prevention, and adaptation, on the other. The two approaches also tend to ignore the prominent role that non-state actors play, including that of international environmental non-governmental organizations (NGOs), particularly as they create new political space or niche.

The purpose of this chapter, then, is to characterize the political niche NGOs are creating and the influence they are building in environmental diplomacy.[2] I begin by explicating the implications for environmental diplomacy of the top-down and bottom-up dichotomy[3] and then conceptualize NGOs as independent actors with their own, often unique, bargaining assets. Finally, I argue that NGOs create alternative linkages between the local and the global levels of politics.

TOP-DOWN APPROACHES

Top-down approaches emphasize traditional diplomacy, in which bilateral and multilateral bargaining is the chief instrument for advancing national goals and reaching accords. National interests and the distribution of power are the primary determinants of outcomes. In the environmental arena the implication of these approaches is that major powers must take the lead to solve serious problems that cross national boundaries. The major powers have the carrots and the sticks, they create the dependencies of other states, and only they can overcome the free-rider problems.

Consequently, top-down implies *global management*. The prevailing model is classical conference diplomacy, including its contemporary manifestations in the Antarctic Treaty System, Law of the Sea, the Montreal Protocol and, most recently, the United Nations Conference on Environment and Development (UNCED). Major powers conduct multilateral conferences and write conventions and protocols. Where necessary, they create or reform institutions to manage new problems as they arise. International organizations are often prominent as coordinators and implementers of state intentions.

In the international development context, the top-down approach emphasizes capital and technology. Substantial infusions of money and expertise are needed to correct resource and development problems. Although bilateral and multilateral institutions and national governments are the primary players, local people need to be involved. But, because they do not have the capital or the technology, they must be trained, directed, and funded. Success is defined in terms of specific projects, such as dams or timber plantations or aquaculture ponds – or, more narrowly from the donor's view, in terms of the flow of funds.[4] Environmental concerns enter the calculus as one more entry on the cost side of the equation. And NGOs are seen to act in an advisory role on the sidelines.[5]

Top-down approaches to international environmental problems are easy to criticize as instances of dominance and neocolonialism, as recapitulations of military and economic power plays, as attempts by powerful elites to co-opt movements and capture progressive ideas for regressive ends. Although such charges often have merit, top-down approaches are, nevertheless, necessary components in any overall attempt to address broad-ranging environmental issues. Despite predictions for decades of the demise of the state system, the system still has the most pervasive influence on international affairs in all issue areas. It is still the most effective mechanism for marshalling resources, human and natural. And it will remain so, even with the collapse of the Soviet empire and the struggles of developing countries. States will continue to provide the resources and create the forums to debate and decide major issues. They will continue to monopolize coercive authority within their respective territories. The major powers will continue to dominate the major financial and trade institutions with their proportionally weighted votes. Nevertheless, the top-down approach has serious limitations inherent in the state system and states themselves, in particular, the inability of states to deal with ecological constraints on economic growth.

One limitation of the top-down approach is the presumption that states, especially major powers, will take the lead. Evidence suggests the contrary. Not only are major powers major sources of environmental degradation and resource depletion worldwide, but they do not, on the whole, lead. They are not the first to initiate international environmental measures. As with human rights, they are not even primary actors in setting the environmental agenda. Often as not, they are obstacles to change, not proponents of change – and for good reason, because the traditional concerns of international relations have been military security, trade, and monetary relations.[6] In these contexts, stability and order are paramount for major powers. Thus, even as states experience ever-increasing environmental threats, defenders of the status quo are not likely to take the lead in reducing those threats.[7]

A second limitation is the very concept of power. What, for example, is the relevance of state 'power' when global warming is a function of activities at all levels of industrial organization and when collective action entails governmental cooperation at every level from the local to the international? Or what is state power when communities resist waste facilities or logging operations and effectively thwart national and international policies? In short, power, traditionally defined, tells little about how serious transboundary environmental problems are likely to be solved.[8]

A third problem is in the nature of the actors themselves. Foreign policy has traditionally been conducted by an elite corps of diplomats. In the latter half of the twentieth century, the small elite corps was complemented by and, in some instances, supplanted by, political leaders and bureaucrats as the issues and tasks have expanded beyond power politics to economic and technological affairs. Moreover, certain functions, such as conflict resolution and human rights monitoring, have been assumed by international organizations and private humanitarian and religious groups.[9]

So, although the work of foreign ministries remains essential for conducting the affairs of state, a much more complex picture of diplomacy emerges when one considers the expansion and complexity of issues, global communications, and the involvement of non-state and intergovernmental organizations in the twentieth century. A classical view of diplomacy is especially inadequate in the realm of environmental conflicts. Professional diplomats trained in international law, diplomatic protocol, the art of negotiations, and, above all, improving interstate relations, and, at least for major powers, maintaining the status quo, are not necessarily equipped to deal with urgent environmental problems.

In part, traditional diplomacy is inadequate because environmental and resource issues tend to be technically much more complex than most diplomatic issues, even arms control and multilateral trade negotiations. But complexity is not the entire problem. Environmental issues require analytic processes that transcend mere technical complexity. These issues require, above all, integrative, interdisciplinary, multilevel approaches – what those schooled in diplomatic protocol, classical European power politics, East–West superpower confrontation or trade negotiations are not accustomed to.[10]

To be sure, institutional innovation as documented in the regime literature reveals efforts to respond to urgent problems. But the empirical work of this volume, of Haas, Young and Osherenko, and of others reveals, as much as anything, that what drives much of regime formation and maintenance is not traditional diplomatic activity, as one might find in the dealings of the GATT, World Bank, or NATO.[11] Rather, when it comes to initiating regime formation and articulating regime norms, diplomats tend to take a back seat to epistemic communities, 'individual leaders', and environmental NGOs.[12]

In the international development arena, top-down approaches suffer from their inability to meet local needs. To the extent that international environmental problems are local in their roots or must ultimately be addressed at the local level, foreign assistance programmes funded by industrialized countries and channelled through large donor agencies are unlikely to fine-tune such aid to local needs. In part, this is a problem of scale: large donors tend to promote large projects, projects that are capital intensive and depend on foreign technology. But it is also a problem of distance and cultural ignorance: donors cannot possibly know all that is necessary to fit their projects to local needs.

To the extent that international environmental problems are addressed by top-down approaches, they will be found wanting. In some cases, traditional diplomacy can be modified to incorporate environmental factors. In others, new institutional arrangements will be required. But in many situations, means of connecting transboundary processes with local conditions will be needed. In all, for both analytic and policy purposes, it is necessary to question the exclusive prerogatives of traditional diplomacy when, in practice, environmental dangers elicit governmental responses that are slow at best. In the international environmental arena, 'policy-makers' must be broadly construed. And, to distinguish among them, criteria are needed to identify those actors whose efforts match in some way the true nature of environmental degradation. Under these conditions, key actors are not strictly, or even primarily, professional diplomats.

BOTTOM-UP APPROACHES

Bottom-up approaches differ from top-down in their emphasis on community organizing, grass-roots movements, local participation, and local decision making. Success is measured less by products – treaties or public works projects – than by processes, especially those that lead to durable institutions that respond to and promote locally desirable solutions to resource problems.[13] The strength of bottom-up approaches lies in their ability to encourage locally tailored responses to meet local needs.

In practice, however, bottom-up approaches suffer from several weaknesses. One is that a scattering of local projects, however successful individually, is not likely to meet in the aggregate the magnitude of regional, let alone global challenges. Decentralization may facilitate local responses, but without a strong multiplier effect they are unlikely to add up to significant societal change. The second weakness is that such projects do little to arrest the larger economic and political forces that compel, say, poaching or unsustainable logging. In short, decentralized approaches by themselves do not make the necessary connections, either laterally or vertically.

In the aggregate, over time, bottom-up approaches may become major forces, just as the citizens movements in Eastern Europe and the former Soviet Union were instrumental in bringing to an end the Cold War. But in the environmental arena with synergistic and threshold effects, there is little assurance that the magnitude or rates of aggregated change will correspond meaningfully to the rates of biophysical change.

Analytically, bottom-up approaches are more intractable than top-down approaches. Organizations, networks, and projects are diffuse and often ephemeral. There is no single hook, no structure, no 'system' analogous to a balance of power or a hegemonic system, upon which to hang one's analysis, let alone to design a policy. Consequently, it is easy to ignore or downplay such efforts. From both the analytic and policy perspectives, it is easier to concentrate on institutional design and institutional reform, to note the contribution of interest groups, and to concentrate on modifying traditional diplomacy to accommodate the needs of interested actors.

The theoretical task, then, is both to make these bottom-up efforts analytically tractable and to account for the necessity of top-down approaches. One vehicle for such an analysis is international environmental NGOs and their role in forging links between the two levels. The diplomatic niche NGOs occupy is neither strictly bottom-up nor top-down. For example, contrary to their promotional claims, neither Worldwide Fund for Nature (WWF) nor Greenpeace are primarily grass-roots organizations. Their employees are not part of the local communities they serve, nor do they share the socioeconomic standards of the poor or working-class people they often attempt to reach. Greenpeace identifies a problem area, enters for a direct action protest, gets the media coverage, and then disappears.[14] WWF funds a conservation project, sends technical advisors, and tries to make the project self-sustaining.

Neither are such groups agents of top-down management. As non-governmental groups, quite obviously they cannot dictate terms to anyone. They cannot tax or legislate or adjudicate. They cannot set foreign assistance policies. They can, however, have influence. As it turns out, that influence is quite unlike that assumed to be essential by traditional analysts and policy-makers. And it is exerted neither from the bottom up, nor from the top down. Rather, where environmental politics is more than the relations of states, NGO influence is exerted by linking the local to the international levels of politics. To make these linkages, environmental NGOs learn to offer to other actors what those actors cannot achieve by themselves. In other words, NGOs become independent actors on the international scene when mutual dependencies arise among key actors, including governments at all levels. Before explicating the nature of the local–global linkages, I first turn to the NGOs' sources of bargaining leverage.

NGO RELATIONS AS AN EXCHANGE PROCESS

NGO bargaining leverage is not built on the traditional power resources of territory and armies. Some NGOs can, however, wield enough economic clout to change governments' or other NGOs' behaviour. For example, in 1991, WWF-US contributed $12.9 million to 407 projects in sixty-three countries. From the early 1980s to the early 1990s, WWF contributed $62.5 million to more than 2,000 projects worldwide.[15] Few developing countries – whether agencies or grass-roots organizations – can ignore such sums of money.[16] Possibly most significant, the World Bank, United Nations Development Programme, United States AID, and other multilateral and bilateral foreign assistance agencies and foundations are increasingly routing their funds through NGOs, both Northern and Southern. In the late 1980s, between 10 and 15 per cent of development assistance funds generated by the OECD member countries were channelled through Northern NGOs.[17] The clout this brings to NGOs *vis-à-vis* both aid recipients and agencies can be significant.[18]

Major international NGOs can command media attention on some issues in ways that few other actors can. Greenpeace, for example, is a master at drawing attention with its publicity stunts, mass mailings, and local organizing. When a local group – whether in Germany or Chile – needs to publicize an issue, Greenpeace has the resources and media savvy to move in.[19] Similarly, WWF can launch a worldwide membership and media campaign when conditions are urgent and, in the process, gain more attention than an international secretariat or a national resource agency.

NGOs also promote communication and muster support, or opposition, for environmental policies. On a given issue, NGOs can reach concerned constituencies that many governments may be hard pressed to reach through their usual press outlets. They can also coordinate lobbying through these networks. For example, in 1987, in anticipation of a key World Bank vote on an Amazon development project, environmentalists in the United States, Europe, and Australia researched and wrote letters jointly to present to their respective governments. Within one week, the coordinated and simultaneous lobbying helped cancel the project.[20] In other cases NGOs can use their reach to concerned constituencies to rally support for state policies.

NGOs also provide scientific and earth-centred knowledge via their own research and their ties with the scientific and land-based, often indigenous or agricultural, communities. Because governments and international organizations tend not to acquire such information routinely, and because their responses to environmental problems are often reactive and crisis-driven, the ready availability of such information is valuable when they do act.[21]

Access to funds, attracting media attention, promoting communications and providing relevant information are the visible manifestations of NGO bargaining assets.[22] Central to NGO assets, however, are qualities of legitimacy, transparency, and transnationalism. Relative to actors in the governmental and business sectors, in the environmental realm, NGOs are perceived as defenders of values that governments and corporations are all too willing to compromise. A community that has had to fight a waste facility siting effort, or a developing country that has tried to ban the importation of hazardous waste, often finds that the corporate and government proponents of such plans represent limited – or distant – interests. If vulnerable people want information and support, they discover they must turn to non-governmental groups, scientific and environmental.[23]

Evidence of the legitimacy of environmental NGOs can also be found in the statements and practices of diplomats. At the Preparatory Committee (PrepCom) negotiations for UNCED, for example, the United States delegation in public and private sessions regularly offered as support for its position the consultation with and the endorsements of US environmental NGOs. It is difficult to conceive of such references in the economic or security realms. Diplomats do not need to validate their trade or arms control positions in terms of interest group support. Clearly, then, governments need NGOs to legitimate their positions and not, contrary to the position of the UNCED leadership, merely to disseminate information.

NGOs' legitimacy derives in part from their mostly single-issue focus and their no-compromise position on environmental matters. Few governmental decision-makers can deal with only a single issue area – they must accommodate a wide range of interests. Nevertheless, on many issues, such as ozone depletion, soil loss, and aquifer drawdown, there are, ultimately, *no* compromise positions. For example, despite the liberal use by US negotiator Richard Benedick of the terms 'balance' and 'compromise' in his description of the Montreal Protocol negotiations, there was only one acceptable rate of ozone depletion – zero.[24] Thus, it is precisely because people do not accept compromise on questions of health or livelihood – risk and cost-benefit analysis notwithstanding – that they turn to environmental groups.

Put differently, people appreciate Greenpeace stunts not just because they are daring and spectacular, or Wangari Maathai's tree planting and defiance of Kenyan authority because she is rebellious. People appreciate such acts because those doing it are risking arrest, injury, and even death – and all for a cause that transcends narrow self-interest. These people are heroes in an age of very few heroes. And they speak to widespread concerns, from pollution to dolphin slaughter, from sanitation to agroforestry.

A second basis of NGO leverage is the ability to enhance the transparency of dominant actors – states, intergovernmental organizations, and transnational corporations. Where asymmetries in information contribute to environmental degradation, NGOs can help rectify the imbalance, thus enabling other actors to organize, resist, and negotiate. For example, when transnational waste haulers exploited the ignorance of governmental officials and private individuals in Nigeria to dispose of hazardous waste from Italy, NGOs in Italy alerted international NGOs, who exposed the entire affair. In the end, the waste was returned to Italy.[25] Similarly, the Pesticides Action Network (PAN) monitors transnational agrochemical companies for compliance with their own code of conduct. PAN's reports on irregularities in trade practices have enabled local communities and workers to work for improvements.[26]

The third basis of NGO assets is the transnational character of NGOs. NGO constituencies can fit the biogeographic scale of ecological problems. As a result, NGOs do not have to be constrained by limited notions of national interest or state sovereignty. For example, in North America, the NGO coalition, Great Lakes United (GLU), has a directorate with Canadian, American, and indigenous peoples representatives such that, when they speak for GLU, they speak for the Great Lakes basin, for the entire ecosystem, not for their respective nationalities (see Chapter 4). Similarly, in Antarctica, NGOs are not bound to the claims of their respective governments, and can organize for the entire region as Greenpeace and ASOC have done (see Chapter 6). At the global level, the various climate-change coalitions transcend national orientations and even North–South differences to address global issues. Under such conditions, NGOs can credibly demonstrate that their interests are broader than that of national representatives. Their allegiance is – or can be – foremost to the ecosystem and to the relevant management processes. Moreover, by representing the broader ecological interest, NGOs can reach out transnationally to a wide range of participants in international forums, governmental and non-governmental, capital city and grass-roots, North and South.

Environmental NGOs gain influence by building assets based on legitimacy, transparency, and transnationalism, assets that, in the environmental realm, states, intergovernmental organizations, and profit-making organizations are hard pressed to match.[27] If international environmental NGOs command these assets what, then, do they seek in return? What exactly is the 'bargain' that defines their political niche?

The NGO bargain

NGOs use their bargaining leverage first to gain access to decision making – governmental, intergovernmental, and corporate – and second to engage directly in the formation and reform of international institutions. Although US-based groups are accustomed to ready access to legislators, such is not the case in most countries, even those with representative governments. More importantly, at the international level, few groups have ready access to international bodies that deal directly or indirectly with environmental issues. The United Nations has accredited NGOs since its inception. But, although access to the General Assembly and conferences like UNCED is important, the international organizations with the greatest environmental impacts – GATT, World Bank, IMF, FAO[28] – grant little or no access to non-members. And protest from the outside has proven largely ineffective. Aufderheide and Rich observe that what protest groups and think tanks alike have failed to appreciate sufficiently is that international development issues are intrinsically political. Bad policies do not remain in effect simply because of an absence of good ideas but also because powerful forces block the adoption of those ideas.[29]

The alternative, Aufderheide and Rich have found, is building NGO coalitions North and South, capital city and grass-roots, to win the support of the political forces that influence the development banks. This approach involves little compromise of basic principles, nor is it non-confrontational. Rather, it recognizes the political underpinnings of multilateral development aid, by bringing 'pressure to bear on these institutions at their most vulnerable points', especially to demonstrate that 'continued ecological neglect could somehow be made to pose a threat to the growth and even survival of the banks.'[30]

Bringing political pressure to bear involves at least indirect access. But direct access to international negotiations – that is, getting to the table when the rules and principles for future international behaviour are being established – may be equally, if not more, important.[31] In negotiations that have few precedents, little predetermined structure, an ill-defined agenda, and fuzzy outcome expectations, simply sitting at the table confers influence. These conditions, arguably, prevail in environmental diplomacy. In the renegotiation of the Great Lakes Water Quality Agreement, the NGO representatives who sat at the table had considerable influence because all concerned were experimenting with an attempt to institutionalize transboundary, ecosystem management practices (Chapter 4). In the attempt to save plant and animal species threatened by trade, the International Union for the Conservation of Nature (IUCN) and WWF drafted the treaty that became CITES and over the succeeding years have been major players in implementing that treaty (Chapter 5). In the Antarctic minerals negotiations, although NGOs did not sit at the table, they did play a prominent role in all other aspects of the negotiations. In addition, NGOs (especially Greenpeace) exerted influence disproportional to their non-state status, by setting up a legitimate scientific research base, which would otherwise qualify states to join the 'exclusive club' of the Antarctic Treaty System (Chapter 6).[32]

In such 'prenegotiations', those who are 'first in' exert influence. Those with prepared position papers and carefully worded proposals for agreement will have influence beyond their structural position in the negotiations. In this regard, an NGO representative can sit comparably – if not equally – with a representative of a superpower or multinational corporation. Moreover, if they have additional assets to offer – expertise, grass-roots support, a transnational base or network, the ability to rectify information imbalances, and, above all, public legitimacy – they carry even more weight; they can actually bargain.

The preceding discussion suggests that, unlike bargaining in the ongoing rounds of GATT or in repeated instances of IMF structural adjustment, much of environmental negotiating is first-time, possibly one-shot, and experimental. It is necessarily reactive in the face of rapid biophysical and social change. In fact, a distinguishing feature of international environmental policies appears to be that they are in constant flux. No sooner is a regime established to protect species threatened by trade, then new arrangements are needed to protect species threatened by deforestation and loss of wetlands. As soon as chemicals known to deplete stratospheric ozone are banned, their substitutes are found to have similar properties (if not other damaging effects). In times of 'turbulence',[33] static notions of order and stability give way, necessarily in the environmental realm, to anticipation and adaptation. In such a diplomatic environment, then, opportunities exist for more than formal, interstate diplomatic exchange. Environmental NGOs do not just demand a presence, they bring assets to exchange in pursuit of their interests.

LINKING THE LOCAL AND THE GLOBAL

NGO bargaining with other international actors, then, is one dimension of the NGO niche in international environmental politics. A second dimension is spatial. That is, NGOs position themselves within both top-down and bottom-up approaches to international environmental policy-making by attempting to link local needs with the challenges of the global ecological crisis.[34] To understand why this linkage is critical, one must be able to explain why, when local needs are relatively immaterial in the security and economic realms – and, hence, of little concern to diplomats – those needs are so important in the environmental realm.[35] This question has not received the amount of theoretical, let alone empirical, attention it deserves.[36] It may, however, be the critical question to answer in devising adequate responses to the crisis. At a minimum, an answer must be grounded in biophysical, institutional, and economic terms.

To illustrate the need for biophysical grounding, the ecosystem management approach to Great Lakes water quality grew out of a frustration with single-media, single-species approaches. Only an approach that was both basin-wide (and, hence, the critical role of the International Joint Commission) and local (for implementation) would succeed (see Chapter 4). Institutionally, as Ostrom has shown, durable common property resource regimes appear to have common features, most of which have a local component.[37] And, economically, as argued later, businesses are less likely to externalize their costs to the extent that production and consumption decisions are localized (as opposed to globalized).

NGOs link the local and the global in part by permeating national boundaries where states and international organizations are reluctant to do so. States typically resist challenges to national sovereignty in questions of security, trade, and foreign investment. And such resistance may become increasingly prominent for natural resource questions, as pressures for exploitation mount. For example, the Brazilian military objected strenuously to the creation of a Yanomami reserve on Brazil's border to connect with a similar reserve in Venezuela, claiming it would weaken Brazil's territorial claims and deny Brazil its right to exploit its resources.[38] But to the extent that transboundary, regional, and even global ecological problems require local solutions, states will increasingly find their sovereignty constrained.[39] States that cannot mediate effectively between international needs (trade and investment, say) and their own local needs (especially, development of sustainable economies) must accept, indeed, may welcome, transnational activities that enhance that mediation.[40] Encouragement by developing countries of such activities is familiar in the context of foreign aid and investment. To the extent that environmental issues are equated with development issues ('sustainable development'), encouraging transnational links is a logical extension, especially when it brings in foreign exchange.

A debt-for-nature swap is an obvious case. International environmental NGOs buy national debt on the secondary market, exchange it for local currencies, and then carry out conservation projects, usually with local NGOs. In principle, the state benefits by retiring a portion (albeit a very small portion) of its debt and, at the same time, funds its resource conservation projects.[41]

States may also find transnational penetration attractive when illicit activity abounds and governments are unable to curb it. In the Philippines, for example, the federal government was unable or unwilling to stop the illegal export of timber from Palawan Island to Japan. Only when a prominent Filipino NGO joined forces with a Japanese NGO and brought a suit against the Japanese government did the practice come to a halt.[42] Presumably, the Philippine government benefited by eliminating illicit activities and by gaining revenues from the trade that was then diverted to legal – and taxable – channels.

Transnational NGO activities can, thus, reinforce established state practices. But when these activities empower local communities or agencies whose work is otherwise overshadowed by security or development agencies, they become, in fact, parallel to, if not subversive of, state practices. Thus, Migdal's notion of a web of social organizations takes on new meaning, as external penetration comes not only from aggressors, multinational corporations, and assistance agencies, but from groups who can fundamentally transform the society from the ground up.[43] When the Environmental Defense Fund (EDF) brought the Brazilian rubber tapper, Chico Mendes, to testify before the US Congress, EDF did more than influence congressional and World Bank votes. The move also sent a signal to the Brazilian authorities that the international community was supporting the rubber tappers' right to organize.

Local–global environmental linkages can be distinguished from traditional linkages in international economic relations by the nature of the actors and the long-term consequences of their actions. Conventional linkages between the international and the local are primarily contractual business relations. Private profit-making units operate at the international level and the local levels with governments and intergovernmental agencies serving as intermediaries.[44] Depending on terms, the contracting parties and the intermediaries alike benefit. If externalities exist – whether immediate through pollution, or long-term through over-exploitation – the contracting parties and governments can ignore or defer them. When, however, those externalities begin to reach home – that is, when they directly affect the financial assets or social conditions of the contracting parties – or when those burdened by the externalities (third parties) gain a voice in resisting the resource and human exploitation, a different dynamic arises.

When the costs of business can no longer be externalized, interested parties – contracting, intermediary, and third party – begin to search for alternatives. Although the primary beneficiaries may resist change for as long as possible, those with the most to gain and the least to lose from change – that is, third parties and, in some cases, intermediaries – are quicker to seek solutions. One tempting solution is to break all links, to withdraw entirely from the global economy when that economy has destructive and inequitable local effects. Assuming unavoidably increasing interconnectedness, economically and culturally, withdrawal is probably a stop-gap measure at best. Moreover, most subnational groups do not have the means to delink. The forest people of Sarawak, for example, cannot declare independence and shut their borders to loggers. But they can seek linkages to the international community, linkages other than those through national or international development agencies. Such strategies do not isolate these people but enable them to make linkages to international institutions – governmental and non-governmental – that help ensure their preferred way of life.

Thus, the alternative to delinkage is the construction of new linkages. Because new linkages threaten prevailing relationships, states and intergovernmental agencies are unlikely to promote them. Such efforts are too easily construed as restrictions on trade, or affronts to diplomatic protocol. NGOs, however, are likely to be agents of such linkages as, by their very constitution, they need pay little heed to national boundaries (although they must, certainly, to national sensitivities).

New linkages may encompass a wide range of activities, from on-the-ground projects to lobbying of governments, from monitoring institutional performance to facilitating foreign assistance negotiations, from regulating international trade to protest and direct action. They encompass relations with a wide range of actors from major powers to small states, from global, regional, and binational intergovernmental organizations to supranational organizations, from multinational corporations to producer cooperatives, and from well-financed, well-connected international NGOs to tiny, ephemeral grass-roots groups.

One way to characterize these linkages is in terms of *upstreaming*. After years of organizing urban and rural communities, grass-roots NGOs North and South are increasingly concluding that the underlying sources of poverty and environmental degradation come not from local inadequacies, but from forces higher up the economic and political ladder, hence, the need to 'upstream.'[45] To solve local problems they are realizing that they must address global phenomena, including the full range of Bretton Woods and United Nations institutions.[46]

Finally, these linkages create their own tensions and conflicts. As a series of bargained exchanges, tensions arise out of the process of determining appropriate exchange rates. For example, if WWF acts as a conduit for conservation funds between the World Bank and a local community, it must negotiate in both directions the terms of the aid flow. Moreover, what is often taken as common purpose and shared interest – say, saving species or reversing ozone depletion – is not necessarily common ground for all parties. The multilateral bank is looking for more reliable means of aid dispersal; the northern NGO is seeking a global environmental impact; and the southern NGO is looking for direct development assistance.

CONCLUSION

Implicit in every prescribed approach to addressing the global ecological crisis is an actor, in particular, an intervening agent. The top-down approach implies great powers; the bottom-up approach implies grass-roots organizations. I have argued that, to the extent that global environmental problems are at once global and local, interventions must occur at all levels and must cross levels. A key set of actors for intervening at both levels and, most importantly, intervening to create new linkages between the local and the global, is international environmental NGOs. States and their international organizations often cannot do this because such linkages threaten existing links necessary to support the international political and economic system.

A fuller understanding of how NGOs fill a growing diplomatic niche requires an understanding of NGO influence. Clearly, it is built not on territory and natural resources nor on the ability to gather taxes and marshall armies. Rather, it is the influence achieved by building expertise in areas diplomats tend to ignore and by revealing information economic interests tend to withhold. It is influence grounded in immediate community needs. From the perspective of leadership, it is influence gained from speaking when others will not speak, from espousing something more than narrow self-interest, from sacrificing personal gain for broader goals, from giving voice to those who otherwise do not have it, from rejecting pessimism and looking for signs of hope. Put differently, it is the influence gained by filling a niche that other international actors are ill-equipped to fill. Moreover, it is influence gained when other actors need what only environmental NGOs can offer. As a result, it is best not to view international environmental NGOs as self-sacrificing altruists. Rather, they should be viewed as actors with their own organizational imperatives, their own frailties, and, most importantly, their own bargaining assets.

Unencumbered by territory to defend or treaties to uphold, international environmental NGOs can trade on these assets for access. With access at the top and bottom, NGOs can create linkages between the local and the global, linkages that respect local needs for sustainable economies and, at the same time, linkages that do not attempt to disconnect localities from, nor avoid confrontations with, the larger forces of global society.

NOTES

1 Gordon A. Craig, and Alexander L. George, *Force and Statecraft: Diplomatic Problems of Our Time* (New York: Oxford University Press, 1983/1990).
2 Clarification of terms is useful at this point. I use *international* not to denote interstate behaviour, but to distinguish from the ever more popular 'global.' In international I include global, regional, cross boundary (two states), and inter-societal (e.g., native culture confronting western culture within a state). For *policy* I adopt Caldwell's definition for the international environmental arena: 'the decisions, agreements and behaviours of peoples, governments, and international organizations relating to human interactions with the biogeochemical systems of the planet' (Lynton K. Caldwell, *Between Two Worlds: Science, the Environmental Movement, and Policy Choice* [Cambridge: Cambridge University Press, 1990, preface, x]).

   Logically, then, diplomacy is not just the traditional conduct of the affairs of states, but a range of relationships conducted by state and non-state actors that involves negotiation and institution building.
3 For useful distinctions between top-down and bottom-up approaches in the domestic public policy arena, see Paul A. Sabatier, 'Top-Down and Bottom-Up Approaches to Implementation Research: a Critical Analysis and Suggested Synthesis', *Journal of Public Policy*, 6 (1986) 21–48.
4 For critiques of traditional international development practices and their impli-cations for environmental degradation, see John O. Browder, 'Development Alternatives for Tropical Rain Forests', in Jeffrey H. Leonard, *Environment and the Poor* (Washington, DC: Overseas Development Council, 1989); David C. Korten, *Getting to the 21st Century: Voluntary Action and the Global Agenda* (West Hartford, CT: Kumarian Press, 1990); John P. Lewis and contributors, *Strengthening the Poor: What Have We Learned?* (New Brunswick, NJ: Trans-action Books, 1988).
5 Ann Hawkins, 'Contested Ground: International Environmentalism and Global Climate Change' (paper presented at the International Studies Association annual meeting, Atlanta, GA, USA, 1992), 10.
6 One can find instances to refute this assertion, the most notable of which would be the US role in the Montreal Protocol negotiations. Even here, however, the driving forces were not so much diplomatic skill, foresight, and global leadership – contrary to the explanation given by Richard E. Benedick, *Ozone Diplomacy: New Directions in Safeguarding the Planet* (Cambridge, MA: Harvard University Press, 1991). Rather, industry pressure – or, maybe better, acquiescence – drove much of the critical bargaining. That pressure was in reaction to the uneven playing field created first by the patchwork of local regulations in the US and then by shifts in the international business climate.

7 Ronnie Lipschutz, for example, argues that states do not have the requisite extended time frames, nor the ability to link cause and effect. When knowledge-based actors such as environmental groups step in, the result is 'not so much "sovereignty at bay" as sovereignty parsed and redefined'; see 'Global Change and Cooperation in the Implementation of International Environmental Agreements: From Theory to Practice', 1991, typescript), 15.

8 Oran R. Young and Gail Osherenko conclude from their analysis of five case histories of environmental regime formation that 'none of the cases offers strong support for ideas derived from the theory of hegemonic stability [and that although] power was important in each of the cases, it is more easily understood . . .as a source of bargaining strength'; see *Polar Politics: Creating International Environmental Regimes* (Ithaca, NY: Cornell University Press, 1993).

9 On changes this century in diplomatic practice, especially in the security and trade realms, see Craig and George, *Force and Statecraft*; Gilbert R. Winham 'Negotiation as a Management Process', *World Politics* 30 (1977): 87–114. For evidence of the increasingly important role non-state actors like the Vatican and the Quakers play in international peacemaking efforts, see Thomas Princen, *Intermediaries in International Conflict* (Princeton, NJ: Princeton University Press, 1992).

10 See the conclusion, Chapter 8, for a more extensive discussion of the inadequacies of traditional politics in the face of novel biophysical threats, especially those characterized by irreversibility and non-substitutability.

11 Peter M. Haas, *Saving the Mediterranean: The Politics of International Environmental Cooperation*, New York: Columbia University Press, 1990); Young and Osherenko, *Polar Politics*.

12 Ethan Nadelmann finds a common pattern historically in the initiation of 'prohibition regimes'. Whereas states condoned, even supported, practices such as piracy, drug trading, and slavery, it was so-called moral entrepreneurs who played a critical role in creating global norms that led to international sanctions against such practices; see 'Global Prohibition Regimes: The Evolution of Norms in International Society', *International Organization* 44 (4) (1990) 479–526. For an application of Nadelmann's findings to issues of international wildlife conservation, see Chapter 5.

13 For a provocative analysis of the need for voluntary, grass-roots organizing see Korten, *Getting to the 21st Century.*

14 This characterization of Greenpeace does not apply to local chapters that do organize and build constituencies. I am primarily referring to the work Greenpeace does internationally. Moreover, as I discuss further later, some evidence exists that these roles are changing. For evidence of the distance Greenpeace has come from the grass roots, see Bob Ostertag, 'Greenpeace Takes Over the World', *Mother Jones* (March/April, 1991) 85; Joy Meeker, 'Greenpeace' (Syracuse University, 1991, typescript).

15 World Wildlife Fund, annual report, 1991.

16 Of course, as donors, NGOs can face many of the same foreign assistance problems experienced by the major donor countries and agencies. Funds go to unintended projects or uses, or simply disappear through corrupt practices. See Thomas W. Dichter, 'The Changing World of Northern NGOs', in Lewis, *Strengthening the Poor* (1988) 177–88. It should be pointed out that the dependencies highlighted here are not all mutual. In fact, a tension NGOs typically face is between their espoused principles of empowerment and their projects that, in practice, breed one-way dependency. Governments tend to be less concerned with such dependencies because foreign aid serves important foreign policy objectives, of which empowerment and self-reliance are rarely a part. But NGOs' legitimacy depends to a large extent on their ability to enhance independent

institutional capacity. NGOs know full well that the safety of a species or habitat is not assured when the community depends on continued external support. The big NGOs like WWF know they cannot adequately address conservation issues worldwide by multiplying projects. They realize they must reach out to the broader forces and overarching institutions – for example, trade and finance – to respond adequately to environmental crises. This argument, then, is yet another reason why NGOs necessarily must link the local with the global.

17 Dichter, 'Changing World of Northern NGOs', 178.

18 In Kenya, this routing of funds through NGOs prompted the government to draft a law that would have restricted NGO activities and siphoned off much of their foreign revenues. The law was withdrawn, however, when NGOs and donors complained. Personal communication, Calestous Juma, director, African Centre for Technology Studies, Kenya, 1991.

19 Greenpeace can also generate resentment, not just from the local business community but from the local environmental groups which are upstaged by the high-visibility events and then, as is often the complaint, left to themselves when Greenpeace departs.

20 *NetNews*, newsletter of electronic information network, San Francisco: Institute for Global Communications, 6 (1) (January, 1992).

21 For an early treatment of knowledge as a bargaining asset in traditional diplomacy, see Klaus Knorr, *The Power of Nations: The Political Economy of International Relations* (New York: Basic Books, 1975). For more recent discussion of knowledge as a means of influence, especially among non-state actors, see the work on epistemic communities in Peter M. Haas, ed., *Knowledge, Power, and International Policy Coordination* (special edition), *International Organization* 46 (1) (Winter, 1992).

22 Because in this book we avoid a strictly functionalist approach to explaining NGO relations, these manifestations are only a partial list of what NGOs do. The purpose here is to use these examples to suggest the bases of the full range of NGO activities.

23 As Hawkins, 'Contested Ground', points out, legitimacy must be considered carefully. Not all actors within the international system, state and non-state, perceive modern environmentalism as legitimate. See also David Barkin and Steve Mumme, 'Environmentalists Abroad: Ethical and Policy Implications of Environmental Non-Governmental Organizations in the Third World', paper prepared for the Third Congress of the International Development Ethics Association, Tegucigalpa, Honduras, 21–8 June 1992; Hira Jhamtani, 'The Imperialism of Northern NGOs', *Earth Island Journal* (San Francisco, CA: Earth Island Institute) 7 (10) (June 1992); Nancy Lee Peluso, 'Coercing Conservation: The Politics of State Resource Control', paper presented at 1992 annual meeting of the International Studies Association, Atlanta, GA, USA. My point here, however, is that, relative to the dominant actors – governments, especially centralized governments, and businesses, especially large corporations with little accountability to the communities in which they operate – environmental NGOs enjoy a higher level of legitimacy.

24 Benedick, *Ozone Diplomacy*.

25 Jennifer Olsen, 'Italian Hazardous Waste in Koko, Nigeria', paper, 1993, University of Michigan, Ann Arbor, USA.

26 Karen Mingst, 'Implementing International Environmental Treaties: The Role of NGOs', paper presented at the annual meeting of the International Studies Association, Mexico, 1993. Mingst uses the notion of transparency in the context of treaty compliance. Yet as the hazardous waste and pesticide examples indicate, the more general case, one that obtains from a transnational perspective of environmental policy-making, is that in which information regarding the full

range of international activities are considered. In fact, given the general consensus that the some 200 extant environmental treaties have little impact, transparency may be most significant with respect to economic arrangements including foreign direct investment and trade. For elaboration of the importance of a transnational perspective, see Chapter 8.

27 A lengthier analysis would differentiate government and business to reveal components in each that, indeed, do represent the interests of those affected by environmental degradation. For example, the US delegation to UNCED was instructed to endure the process with minimal political cost. The interagency task force created to support the delegation did, at the level below political appointees, create effective intergovernmental communication that, under a more proactive administration, could have contributed positively to the objectives of UNCED (confidential communication with task force member, 1992).

28 The General Agreement on Tariffs and Trade; the International Monetary Fund, and the United Nations Food and Agriculture Organization.

29 Pat Aufderheide and Bruce Rich, 'Environmental Reform and the Multilateral Banks', *World Policy Journal*, 5 (2) (1988) 305–6.

30 Aufderheide and Rich, 'Environmental Reform and the Multilateral Banks', 308.

31 See I. William Zartman, 'Prenegotiation: Phases and Functions', in *Getting to the Table: The Processes of International Prenegotiation*, ed. Janet Gross Stein (Baltimore, MD: Johns Hopkins University Press, 1989).

32 In the trilateral negotiations over the North American Free Trade Agreement, United States NGOs that cooperated with the US administration and helped draft and promote the agreement were rewarded. For example, the president of the National Audubon Society said in a news interview that his NGO's support was partly based on an assurance from the Administration that an agreement on migratory birds in the Western Hemisphere would be added to a list of environmental pacts protected from legal challenges by the free-trade agreement. Keith Bradsher, 'Side Agreements to Trade Accord Vary in Ambition', *New York Times*, September 19, 1993, 1, 15.

  The six major environmental NGOs thus appear to have had a major impact in the formation of the regional trade regime – at least with respect to strictly defined environmental issues – and they did so by trading their support for specific provisions.

33 James Rosenau, *Turbulence in World Politics* (Princeton, NJ: Princeton University Press, 1990).

34 NGOs are not the only ones to make these links. Intergovernmental agencies such as UNEP, scientific organizations such as the International Council of Scientific Unions, and multinational corporations certainly are prominent in this regard. The focus on environmental NGOs is justified, however, because these groups have a primary commitment to environmental issues. Intergovernmental organizations, by contrast, are primarily committed to their member states and the state system they serve; scientific groups to scientific research; and corporations to their shareholders. Just as religious organizations have a greater claim on a correct moral stance on a given issue, environmental groups, by virtue of their single-minded focus, can claim a privileged stance (ecologically, even economically, if not necessarily morally) on environmental issues. This is not to say they are always right, or that they eschew exaggerated claims or that they are totally uncompromising. In fact, as discussed previously, much of what they do can be best understood as bargaining, as making trades on their assets for what only governments and corporations can offer. They are, in short, international political actors but, on environmental issues, they carry more credibility than other actors whose functions and allegiances are centred elsewhere.

35 Economic dislocation as a function of trade patterns is, of course, of great concern in the economic realm. But in the current trade regime governed by principles of free trade and non-discrimination, localized negative impact is cause for creative side-payments, not an independent source of decline in the gross world product.

36 The few notable exceptions include Rosenau, *Turbulence in World Politics*; Daniel Deudney, 'Global Environmental Rescue and the Emergence of World Domestic Politics' (paper presented at the International Studies Association annual meeting, Vancouver, BC, March 1991); and Lipschutz, 'Global Change and Cooperation.'

37 Elinor Ostrom, *Governing the Commons: The Evolution of Institutions for Collective Action* (Cambridge: Cambridge University Press, 1990). Agricultural economist C. Ford Runge comes to a similar conclusion regarding the importance of local solutions to resource depletion problems in his analysis of the assurance problem: 'Common Property and Collective Action in Economic Development', in Daniel W. Bromley, ed., *Making the Commons Work: Theory, Practice, and Policy* (San Francisco, CA: Institute for Contemporary Studies Press) 17–39.

38 *New York Times*, November 19, 1991.

39 See Oran R. Young, *International Cooperation: Building Regimes for Natural Resources and the Environment* (Ithaca, NY: Cornell University Press, 1989), for discussion in the context of international regimes of how states must limit their sovereignty to gain the advantages of international cooperation.

40 These conditions apply not just to the South. Industrialized economies are not necessarily any more sustainable than less-industrialized economies. In fact, one can surmise that, should major ecological thresholds be reached, the resulting ecological and economic disruption would be relatively more severe in the North.

41 For critical assessments of debt-for-nature programmes and the role of inter-national NGOs, see Susanna Hecht and Alexander Cockburn, *The Fate of the Forest: Developers, Destroyers and Defenders of the Amazon* (New York: Harper Perennial, 1990); and Alan Patterson, 'Debt-for-Nature', *Environment* 32 (10) (1990): 5–13, 31–2.

42 Personal communication, Haribon Foundation, 1991.

43 Joel S. Migdal, *Strong Societies and Weak States: State–Society Relations and State Capabilities in the Third World*, (Princeton, NJ: Princeton University Press, 1988). That any number of non-governmental groups can actually transform societies is, of course, a big claim. In part it is the solution offered for the North by Lester Milbrath, *Envisioning a Sustainable Society: Learning Our Way Out* (Albany, NY: SUNY, 1989), and for the South by Korten, *Getting to the 21st Century*. It is indeed contestable whether a mainstream northern conservation group like the World Wildlife Fund aims to fundamentally transform the societies in which it operates. In recent years, there is some evidence of organizational learning whereby 'human needs', 'capacity building', and 'local solutions' increasingly pervade WWF literature. Such groups may not fully play out the implications of such approaches – especially in full view of their major donors – but those working on the ground appear increasingly to be coming to the conclusion that 'local involvement' is not hiring local people as park wardens, and that collaboration with central governments contributes more to the problem than to the solution. For elaboration of these developments, see Chapter 5 and Thomas Princen, 'Ivory, Conservation, and Transnational Environmental Co-alitions', in Thomas Risse-Kappen, *Bringing Transnational Relations Back In.*

44 Like other analysts grappling with the local–global nexus, I am vague about the term 'local'. From the international relations perspective, the term encompasses all that is less than the state system and states themselves. It includes subnational actors, governmental and non-governmental, which are not tightly linked to the

dominant reward and control system of the state. It may include, quite literally, small communities but it may also include organizations that span large populations, even those that extend beyond national boundaries.

From the perspective of resource institutions, the defining characteristic of 'local' is that it is that social unit that can create durable mechanisms for sustainable use of the resource. In *Governing the Commons*, Ostrom begins to lay out the necessary conditions, at least in small-scale operations with an enabling external political environment. What is not clear from her work, however, is the importance of scale. It can be presumed that certain of her eight principles have scale limits. That is, it is difficult to imagine applying self-monitoring and the right to organize to the global commons of the atmosphere and the oceans. It is, however, imaginable to apply them to resources that are geographically large and that cross political boundaries because, according to Ostrom, the key to durable commons governance is delimitation of the resource, participation of the appropriators in creating, maintaining, and adapting the regime, and the ability to restrict use. These criteria, with their implicit limitation on scale, may offer the best operating definition of 'local'.

45 David Korten, *Getting to the 21st Century: Voluntary Action and the Global Agenda* (West Hartford, CT: Kumarian Press, 1990).

46 One of the lessons of UNCED for many NGOs is that the real action is elsewhere, namely, at the GATT, IMF, and World Bank meetings, where NGOs are not granted the same degree of access. In fact, many NGOs have concluded that the tremendous accommodation of NGOs at UNCED enabled them to do little more than squabble over innocuous Earth Charter and Agenda 21 language. See Chapters 7 and 8.

3   NGOs and transformation: beyond social movement theory

Matthias Finger

The emerging phenomenon of environmental non-governmental organizations (NGOs) in world politics in terms of social movement activism has been transformed from a single concentration in national politics in the 1970s to an increasingly broad focus at global and local levels in the 1980s that relativized the importance they initially attributed to national politics. This change in focus, along with the institutionalization of the green movement, reflects the emergence of international environmental NGOs. The transformation of the green movement must be viewed within the larger context of the much more profound cultural transformation that led from political ecology in the 1970s to global ecology in the late 1980s.

The globalization of ecology and the corresponding transformation and globalization of environmental activism corresponds to the phenomenon some call 'global civil society'[1] or 'turbulence in world politics',[2] paralleled by 'the emergence of a multicentric world consisting of thousands of nonstate, nonsovereign global actors, [which] coexist in a nonhierarchical relationship with the state-centric system.'[3] International environmental NGOs may be the most significant expression of this turbulence. They express the transformation environmental movements themselves have undergone; national actors aiming at national political influence are becoming transnational actors, thus fundamentally changing their relationship with traditional, national, and state-centric politics.

Can the emerging international environmental NGO phenomenon be adequately conceptualized in terms of social movement theory that has evolved from theorizing social movement activism at national levels, particularly in western democracies? If this is the case, one can view environmental NGOs as the continuation, on a global level, of what social movements were (and still are) on a national level. If, however, social movement theory inadequately accounts for the emerging environmental NGO phenomenon, further development of social movement theory will be justified, along with conceptualization and theorization of international environmental NGOs.

Social movement theory, although universal, originates in the sociological theory of collective behaviour and collective action developed by the Chicago school of sociology. As such, social movement theory goes back to psychosociology and the study of individual behaviour within groups. Collective action, according to this theory, can be triggered in various ways, depending essentially upon the theoretical framework to which one refers. One can distinguish three main schools. All of them are fundamentally ahistorical. Indeed, collective action can occur either as a result of relative deprivation,[4] as a strategy to articulate common interests,[5] or as a response to economic or political conflicts. In a political context the purpose of collective action is social change.

Given these historical foundations, social movement theorists have mainly been interested in the promoters of collective action – activists. In this perspective, conflict theorists often seek to identify the main conflicts around which collective action crystallizes and evolves. When characterizing some of these elements, such as the promoters of collective action and the nature of conflicts, social movement theorists observe profound historical changes that suggest new social movements.[6] At times, new social movements are equated with new politics, in particular when conflicts include and evolve around new, especially environmental,[7] cultural,[8] and lifestyle issues.[9]

If old social movements are carried by the working class, new social movements are carried by parts of the new middle class.[10] Empirical observations of emerging new characteristics of country-specific social movements have not generally led to a search for alternative social movement theories. Nor has the emerging transnational phenomenon led experts to examine critically and, perhaps, update their theories. One must view NGOs in relation to the globalization of ecology in general, and the changing relationship to state-centric politics in particular.

SOCIAL MOVEMENT THEORIES: THE NATIONAL LEVEL

Social movement theories dominant in the 1990s are all rooted in the collective social action approach, which can be traced back to the origins of sociological theory. Further theorization by activists in the North during the social activist period in the late 1960s and the 1970s took a political turn, focusing mainly on the movements' political actions and concerns, most of which were limited to the national level.

Cyclical theories: the example of Alain Touraine

Cyclical theories assume that, as with various historical movements, social movements occur for similar reasons and have similar destinies. The conceptual framework that helps to explain the emergence of the labour movement in the nineteenth century, therefore, is equally valid to explain the emergence of the green movement in the early 1970s.

Alain Touraine's[11] theory is, without doubt, the most elaborate example of the theory that analyses movements in terms of political cycles: social movements strive for political power at the national level, but, in the process

of reaching that power they are co-opted. Consequently, new social movements arise. Touraine's theory is typically modelled after the labour movement. New social movements, such as the green movement, are seen as the
equivalent in a post-industrial society of old social movements, such as the
labour movement, in an industrial society.

Touraine's theory is grounded in a materialist world view: linear development of productive forces (industrial development) provides society
with an ever-increasing capacity to act upon itself. The key instrument of this
action is the nation-state. Social movements must, according to Touraine,
take advantage of this ever-increasing capacity of modernizing societies to
act upon themselves by seeking control over the nation-state. In other words,
movements, if they want to be part of the modernization process, must strive
for political power at the nation-state level.

Striving for power is a struggle: it is conflictual, occurring between classes,
oppressors and the oppressed. The objective is participation in state power
but, more generally, the contention is about participation in defining what
Touraine calls historicity, a society's capacity to act upon itself. In this
struggle, the nation-state is considered a neutral instrument, and social
movements must seek to conquer it.

Touraine has heavily influenced social movement theory and provides a
strong conceptual framework that defines what a social movement is and what
it is not. Basically, to deserve the label, a social movement needs to be
struggling to participate in (national) political power. Consequently, Touraine
rules out pressure groups, which only struggle for their interests; anti-
systemic movements, which destroy the very instrument they are supposed
to conquer; national movements, which are prehistoric in the sense that they
fight to establish the nation-state system; and cultural movements, whose
main error is to conceptualize transformation as value change and not as a
political struggle. Had Touraine been aware of the international environmental NGO phenomenon, he would certainly have ruled it out as well.
International NGOs, especially, are not struggling for state power, they are
tugging and pulling at the states.

When defining social movements, Touraine assumes that a continuous,
linear process underlies the development of productive forces, a process that
conveys more and more power to the (national) political system; that all
social forces must struggle to participate in this power at the nation-state
level; and that all movements can be described, therefore, in political, class-
struggle terms. All three assumptions are questionable today, particularly
when considering the globalization of ecology and ecological problems. It is
not obvious whether the development of productive forces actually conveys
increased power to the nation-state. The opposite seems to be true when one
looks at the global environmental consequences of this development: the
nation-state is weakened and/or becomes a problem.[12] Why, then, should
movements struggle to participate in national power?

The application of Touraine's theory to the international environmental

NGO phenomenon, therefore, appears to be rather limited. Some international environmental NGOs have been striving for political power at the national level to use the political system as a means to act upon society, in general, and to solve environmental problems, in particular. International environmental NGOs, however, do much more than that. In Touraine's theory, such NGOs are at best social movement organizations which must, if they want to be efficient, become, sooner or later, political parties. His theory does apply to the political ecology movement of the 1970s and its translation into green parties early in the 1980s – a phenomenon, however, that was limited to some Western European countries. Its geographical limitation, its conceptualization of social movement activism as a purely political form of activism, and its reference to national politics alone ill suits Touraine's theory of social movements to account for international environmental NGOs.

Linear theories of social movements

Linear theories state that social movements, whenever they emerge, are unique, and must be analysed as such. A movement's uniqueness is tied to the fact that the process of industrial development is linear and produces unique societal effects. Social movements are related to the societal effects caused by industrial development. Two of the most influential linear theories are those of Claus Offe and Jürgen Habermas.

The theory of Claus Offe

Claus Offe[13] views social movements, in particular new social movements, as means to help the political system evolve and adjust to the new requirements industrial development places upon it. According to Offe, the political system – the nation-state – is the key to managing the process of industrial development and its societal consequences because the political system is viewed as a regulator between the economic system, on the one hand, and civil society, on the other. Offe implicitly assumes in his theory that, as industrial development progresses, the political system will have to extend its regulatory activities more and more into the economic system as well as into civil society.

Between the political system and the unpoliticized civil society, because of this historically unavoidable process, is some sort of grey zone that is about to become politicized. Social movements emerge, according to Offe, within this field of not-yet-institutionalized politics. They are an expression of this grey zone, and although they often protest the political system's claim on this not-yet-politicized zone, they nevertheless are instrumental in helping the political system lay claim to it and institutionalize it. In short, they are an expression of a transition period. Social movements, therefore, are social actors that help politicize a field that previously belonged entirely to civil society. They thereby help the political system extend its activities into civil society. They help the political system to do something that, because of the process of industrial development, is inevitable anyway.

Offe's view of social movements is, therefore, a functionalist one in which social movements basically have two functions: they contribute to the politicization of civil society, and they help the political system become attentive and adapt to the new challenges brought by industrial development. In other words, social movements help the national political system to adapt, evolve, and, to a certain extent, learn. Offe refers, in particular, to the environmental movement of the 1970s: social movements bring up new issues, politicize them within civil society, and prepare the grounds for the political system to integrate them.

Offe's theory is not necessarily limited to social movements. Many other groups and actors, for example, national and international NGOs, can also function to open up new political space. Many NGOs as well as NGO observers do see their main role as helping governments take up new issues. Offe's theory of social movements, however, is limited to national politics, because political space is always opened in an existing national political system. International environmental NGOs, although they may affect some national political systems, also do many other things. They (re)define politics above, below, and beyond nation-state politics at levels where the national political system cannot simply move in and take over. Offe's approach, therefore, remains limited to the nation-state, a limit international environmental NGOs precisely seek to overcome in practice, and which theorists seek to overcome conceptually.

The theory of Jürgen Habermas

Sociological theories of social movements such as that of Habermas[14] can be traced back to the first generation of critical theorists: Horkheimer, Adorno, Marcuse, and others. Their reference for social movements is the National Socialist movement in Germany in the 1930s. Its emergence and that of fascism, is interpreted by critical theorists as the result of the hegemony of technical rationality (Habermas), technological rationality (Marcuse), or instrumental reason (Horkheimer). If rationality prevails, a society becomes unhealthy because technical rationality is said to destroy a society's capacity to critique and reflect. It alienates people and diminishes a society's ability to learn and master its own evolution and future.

Habermas, like all other critical theorists, sees social movements as both an expression of alienated social reality and a healthy reaction against it. The more the technical rationality invades the life-world, the everyday socio-cultural reality in which individuals live, the greater the chance that citizens will react in a social movement. Yet, the chance is also greater that the citizens' reaction will be irrational. In Habermas's concept, therefore, social movements are basically a healthy yet irrational reaction against the so-called colonization of the citizens' life-world, as well as an attempt to re-establish the autonomy of that life-world. Striving to autonomize the life-world, to liberate it from domination by technical rationality, is a sociopolitical and a socio-cultural struggle not necessarily limited to the traditional political system. Autonomization of the life-world, however, is necessarily accompanied by a strengthening of the political system, which Habermas identifies at the national level.

Rooted in Marxism and its distinction between infrastructure and superstructure, between the modes and the relations of production, Habermas sees the evolution of society in terms of labour, on the one hand, and interaction, on the other – that is, between the evolution of the forms of production and the evolution of the life-world. In between, the political system masters this evolution in both fields, labour and interaction. Moreover, it ensures that technical rationality does not invade the life-world. An overwhelming technical rationality, therefore, is synonymous with a weakening of the political system. If this happens, social movements arise. If they successfully manage to prevent this invasion, they simultaneously restore the autonomy of the political system as the ultimately neutral mediator between labour and interaction.

Habermas's theory is not necessarily limited to social movements; it can easily be applied to NGOs. Nor is his theory necessarily limited to the national level, given its level of abstraction. In practice, however, Habermas's theory remains limited by the fact that the political system he envisions, in which social movements strive to restore the autonomy of the system and of the life-world, is practically a national system.

A much more fundamental critique must be addressed to Habermas's theory that is basically identical to those previously addressed to Touraine and Offe and applies to all social movement theories based on Marxist inspiration. In Habermas's view it is particularly obvious that the primary function of a social movement is to strengthen the political system to restore its autonomy-level does not matter. As explained in this volume and a lot of empirical work, international environmental NGOs only very marginally pursue this function, if at all. Marxist-inspired social movement theories, when applied to the NGO phenomenon, basically miss the point.

Historical theories: resource mobilization

Resource mobilization theory[15] actually does address some of the weaknesses of Marxist-inspired social movement theories, as it combines collective action theory with organizational theory. But the price is the loss of a historical, dynamic perspective on social movements. Resource mobilization theory assumes that it is rational for citizens to participate in the political system, which is simply the steering system of society and not necessarily the nation-state referred to by Marxists. Society is, therefore, basically an aggregate of rational individual actors and not necessarily, as Marxists saw it, a structured mass of (potentially) responsible and autonomous citizens who always remain defined relative to the nation-state. Moreover, society, according to resource mobilization theorists, is made up of multiple organizational structures. Social movements, then, are basically organizations like all others that aggregate rational individuals. The historical origins of resource mobilization theory stem from the conceptualization of consumers' movements and public interest groups in the United States. Resource mobilization theory continues to reflect a typically American approach to public participation; social movements remain simply one means rational actors can choose from if they want to make a difference in the political system.

According to resource mobilization theory, social movements are organizations that help rational actors participate more effectively in the political system than in other kinds of organizations or in purely individual capacities. To help them make a difference, social movement organizations mobilize various resources, for example, skills and values. Such organizations, quite naturally, compete with lobbies and political parties, a competition, however, that is not always fair. Quite logically, the latest version of resource mobilization theory studies the role of so-called political opportunity structures[16] to explain social movements – the mobilization of resources for political participation – as a function of a particular political system. Political opportunity structures is the branch of resource mobilization theory that probably comes closest to the ones inspired by Marxism, because it ultimately defines social movements with respect to and as a function of nation-states and their political structures.

Resource mobilization theory assumes that participation is usually in the national political system. It is difficult, then, to apply the theory successfully to international environmental NGOs, a criticism that is particularly valid for the theory of political opportunity structures. Even if one could stretch resource mobilization theory to view international environmental NGOs as a form of resource mobilization, there is no international system to lobby. National NGOs can, indeed, be captured by resource mobilization theory, but as such they become lobbies. This fundamentally functionalist definition of NGOs, social movements, and, more generally, public participation neglects the political dimensions of social movements highlighted by Habermas, Touraine, and Offe. Resource mobilization theory has a strong bias toward (individual) rational choice, which, in turn, neglects the emotional dimensions of social activism.

Conclusion: the limits of social movement theory

All the social movement theories discussed have a major bias that makes it difficult to use them as models for theorizing NGOs, in general, and international environmental NGOs, in particular. They are biased toward politics at the nation-state level.

Of the theories discussed, resource mobilization theory probably best accounts for the evolution of NGOs in purely descriptive terms. If one applies resource mobilization theory, however, NGOs become lobbies, efficiency-seeking vehicles for participating rational actors. Participation originates at the national level, although it could eventually be extrapolated to the global nation-state or global political system. Touraine's, Offe's, and, to a certain extent, Habermas's theories have a similar bias toward citizens' participation at the national level. Yet international (environmental) NGOs are not primarily concerned with participating in national politics, as the social movement theories discussed so far suggest. They seek, rather, to develop solutions to global and local (environmental) problems, sometimes in collaboration with governments, sometimes against them, often below, above, and beyond traditional nation-state politics by tugging and pulling at states. Unfortunately, none of social movement theories discussed can be used to conceptualize and theorize the other functions that international environmental NGOs fulfil, which might well be at the core of what they are all about.

The bias of dominant social movement theories toward nation-state politics goes hand in hand with another, more profound, bias toward industrial development. All discussed theories implicitly consider industrial development to be inevitable, sustainable, desirable, and even necessary, so that the nation-state can fulfil its political functions. Moreover, all the theories discussed assume that the national political system, the nation-state, is the key actor, a neutral instrument that helps society manage the process of industrial development as well as its consequences. Social movements, then, are conceptualized relative to that neutral instrument. They strive to conquer it (Touraine), or to help it evolve, learn, and extend its influence into not-yet politicized civil society (Offe), or to restore its autonomy as a mediator between labour and interaction, that is, between the infrastructure and the superstructure (Habermas). Resource mobilization theory, especially its latest version concerned with political opportunity structures, is probably the most typical expression of this ultimately national political concept of social movement. Movements are said to mobilize (human, financial, institutional) resources to participate in the national political system, which is seen as a neutral tool to manage industrial development and, thus, help individuals to fulfil their aspirations. For many international environmental NGOs the idea that the nation-state is the ultimate actor, especially in environmental and development arenas, is obsolete. All social movement theories that conceptualize NGOs and similar actors as striving to participate in the management of industrial development at the national level, therefore, fail to account not only for the NGOs' functions today, but also for their nature.

Considering these limitations, are social movement theories at a global level better suited for conceptualizing international environmental NGOs?

SOCIAL MOVEMENT THEORY AT THE GLOBAL LEVEL: THE EXAMPLE OF MARC NERFIN

Since the 1970s many authors have been writing about global social movements: Chadwick Alger, Elise Boulding, Richard Falk, Johan Galtung, André

Gunder-Frank, David Korten, Rajni Kothari, Ashis Nandy, Hilkka Pietilä, Roy Preiswerk, Majid Rahnema, Ignacy Sachs, Dhirubai Sheth, among others. All share a similar analysis of the phenomenon, because they all extrapolate national social movement theory to the phenomenon they see globally. Epistemologically, this view probably comes closest to Claus Offe's conceptualization of social movements. Marc Nerfin's[17] theory of the third system best encapsulates the main epistemological and theoretical elements shared by the authors mentioned.

Third system theory

The point of departure for third system theory lies in the observation that there is a generalized 'development crisis' in the South and in the North. 'This crisis is simultaneously economical, financial, ecological, social, cultural, ideological, and political', explains Nerfin.[18] The way things develop in the 1990s has become, Nerfin argues, a threat to our common security in facing this overall development crisis; third system theorists observe a growing movement that is seeking control over the very process that threatens everybody's security. It is, therefore, a movement of all people who suffer, in one way or another, from the current development crisis, whether economically, socially, culturally, or ecologically. Because it is global, this movement is highly diverse.[19]

Third system activities take various forms, according to Nerfin.[20] He mentions in particular the realization of immediate projects, advocacy, and holding people responsible for their acts and decisions. Third system theorist David Korten[21] sees the citizens movements as playing four critical roles: advocacy, or what he terms catalysing systems change (e.g., redefining policies, transforming institutions, and helping people define, internalize, and actualize a people-centred development vision), system monitoring, protesting that facilitates reconciliation with justice, and implementing development programmes.

The underlying social analysis of third system theory continues to be inspired by Marxism and humanism, as third system theorists perceive a fundamental opposition between the oppressors and the oppressed. This fracture, as Nerfin calls it, is the result of a political problem, the bad management of human affairs:

> The fracture which more and more divides each society is more profound than the traditional gaps, East–West or North–South: the two Indias, the two Chiles, the two Hollands, the two USAs, the two worlds, the one of the powerful, the rich, the ones who have a job, the ones who participate and the powerless, the poor, the unemployed, the disenfranchised, worse, the ones that are not needed any longer as they are not economically useful anymore. This fracture translates into underdevelopment, maldevelopment and other poisoned fruits of the same bad management of human affairs everywhere on this planet.[22]

Quite logically the solution to the problem, as defined by third system theory, has to be sought on a political level, most immediately by focusing on today's politically most relevant actors, i.e. people. In Nerfin's terms:

> In contrast with government power – the Prince – and economic power – the Merchant – there is an immediate and autonomous power, sometimes obvious, always present: the power of the people. Some, among the people, become aware of it, get together, act and become citizens. The citizens and their associations, or movements, when they neither search nor exercise governmental or economic power, constitute the third system. By contributing to make visible what is hidden, the third system is an expression of the people's autonomous power.[23]

The third system equals the currently oppressed people of this planet. They must move out of oppression to become citizens. The term 'third system' reflects this fundamentally emancipatory idea.

The association with the expression 'third world' is more than deliberate: the two expressions stem from the same source; both recall the third estate of the French *ancien régime*. Before the revolution of 1789, French society was divided into three 'estates': noblesse, clergy and the third estate i.e., the majority. Alfred Sauvy was the first one, in 1952, to use the expression 'third world' in referring to the periphery, or South, an expression which has been very successful since. However, 'third system' is conceptually closer to 'third estate' than 'third world' is to the two other expressions. The latter concept is geopolitical: it concerns countries. The two former concepts are sociopolitical: they concern people, and it is people where the third system stems from.[24]

A better management of human affairs is, therefore, achieved, according to Nerfin and other theorists, by third system politics. Third system politics will lead to another development

> oriented towards the satisfaction of all human needs far beyond the 'basic needs'; it is autonomous, endogenous, in harmony with nature and therefore sustainable, and parallels structural transformations increasing peoples power. In other words, another development means that people get organized in order to develop themselves by themselves and for themselves.[25]

In short, third system politics leads to what Korten calls 'people-centred development', characterized by the three following basic principles:

1 Sovereignty resides with the people, the real social actors of positive change.
2 To exercise their sovereignty and assume responsibility for the development of themselves and their communities, the people must control their own resources, have access to relevant information, and have the means to make the officials of government accountable.
3 Those who assist the people with their development must recognize that it

is they who are participating in support of the people's agenda, not the reverse.[26]

Third system theory reduces the environment and development crisis to a political crisis caused mainly by non-participation in the development process. Third system politics, therefore, is about increasing people's participation in decision making at all levels of society. Only through such participation, it is argued, will people become able to ensure their role as sovereign actors.

Critical discussion of third system theory

Third system theory has the potential to overcome the main limitations of social movement theories as one theorizes the nature and the roles of international NGOs. Instead of focusing on citizens' participation in national politics, third system theory concentrates on people as the link between global and local levels. People seek a political expression of this linkage, the NGOs, and bypass national politics. Political action takes place on global as well as local levels. The third system theory links and comes much closer to what international NGOs are about. Third system theorists use the term NGO, not social movement. NGOs are the most typical actors encapsulating this link between global action and the citizens: through NGOs citizens have found a means to express themselves on a global level. Therefore, international environmental NGOs draw their legitimation from citizens who no longer refer to national boundaries.

According to Nerfin,[27] the third system can achieve global relations in at least two ways: through the UN system and by networking. NGOs are more representative than are national governments. Empowering NGOs as relevant actors within the UN system began in the late 1980s.[28] Many observers consider the UNCED process – the process leading up to the United Nations Conference on Environment and Development – as the best example of this approach to global relations. NGO networking, facilitated by modern communication technologies, favours more democratic participation. In short, third system theory is about citizens participating in global decision making. Social movements on a national level and NGOs on a global level share similarities. Social movements functioned as key actors to get citizens' voices heard at the national level; in third system theory NGOs function as global social movement organizations, expressing people's needs and interests, and seeking participation in global decision-making. Because it is people-centred, third system theory considers NGOs to be beyond traditional lines of North–South or East–West conflict. Third system theory defines politics to suit oppressed people's needs and aspirations.

Third system theory is not without some serious epistemological problems, which stem from the fact that it extrapolates a purely political concept of citizen participation in the development process on a national level to the global level. Third system theorists attempt to redefine development (or define another development) from the perspective of the people. Ultimately, third system theory is about global democracy (global domestic politics). But global domestic politics calls for a global system modelled basically after the nation-state. NGOs are, therefore, conceptualized as global social movement organizations, whose main activity, in essence, is to lobby the global political system. I criticize this view from four different, yet related, perspectives: philosophical, cultural, political, and ecological.

Third system theory can be criticized on the same grounds as resource mobilization theory and theories of collective action because it conceptualizes global citizens politics via NGOs as some sort of interest aggregation, thus ignoring such sociological phenomena as institutionalization, power, and control.

Third system theory's point of departure is in a political rather than cultural definition of people. People are defined as individuals, so-called 'world citizens' without cultural roots. Their activities, values, and behaviour are not viewed as shaped by society and culture. As a result, the needs and interests of all oppressed citizens are considered similar, comparable, of equal value, and, therefore, aggregatable.

The global political system that third system theory calls for does not exist yet. Furthermore, such a system probably is not desirable. Conceptualized as a social movement on a national level, third system politics defines itself as a lobbying activity of a basically non-existent global political system, rather than as the innovator inventing new forms of politics.

Third system politics is conceptualized as public participation in global decision making about development, about the distribution of goods produced by development, and about global resources and risk management. The very process of industrial development, in particular the fact that this process is unsustainable, is no concern of third system politics. The concept is purely political.

Overall, third system theory operates within the same conceptual framework as the dominant social movement theories: citizens mobilize to participate in the political process, which has been 'shifted upward' from the domestic to the global (UN) level. The fact that the unit of reference is now some sort of global political system should have changed the terms of reference of the theory: at this global level, activism is not so much about participation and influencing existing structures and decision-making processes but about creating and inventing them. This is what international NGOs – in particular, international environmental NGOs – are really doing. Therefore, NGOs cannot be conceptualized, as is third system theory, as the aggregated collective action of the oppressed on a worldwide scale. In short, third system theory is a too-rapid extrapolation of social movement, in particular, resource mobilization theory, from a national to a global level.

TOWARD A NEW THEORETICAL FRAMEWORK: ENVIRONMENTAL NGOs AS AGENTS OF SOCIAL LEARNING

Among all the social movement theories discussed here, third system theory is certainly the most useful for conceptualizing international environmental NGOs. It is the only social movement theory with a global focus. One can even make the case that NGO is a third system concept; third system theorists generally use the term, whereas social movement theorists at the national level do not. For third system theorists, NGOs are basically lobbyists for a global political system.

Third system theory has shaped today's dominant view of what NGOs are all about. This view, however, is still very narrow; NGOs remain defined in purely political terms. Because their primary identity is political, so are their main roles. But NGOs play economic and cultural roles also, especially in development, and conceptualize and frame issues. A purely political definition of NGOs is much too static: the global political system is more or less taken as given, and the role of NGOs is mainly to participate in the decision-making processes of that system. Although this is certainly part of what international environmental NGOs do, the contribution of social movement theory, in general, and third system theory, in particular, to the conceptualization of NGOs is basically limited to their political dimensions.

This is because social movement and third system theories are still embedded in a traditional concept of politics which refers to a cultural model shaped by modernity and modernization, that is, by the ideas of Enlightenment, rationalization, and continuous industrial development: citizens are to become enlightened participants in a thriving democracy. Social movements and other groups exist to help these citizens and society more generally in order to achieve this goal. They also seek to influence the course and the pace of this evolution by making good use of the political system. Bargaining between citizens and their representatives are necessary means toward this end. The political system remains the point of reference, the actor whose behaviour ultimately needs to be influenced and altered. Without doubt, many activities of environmental NGOs can easily be explained in reference to this framework where traditional politics remains the cultural model. Yet, this model cannot explain all the characteristics of environmental NGOs and all the roles they play. It is true also that the process of modernization of the past one hundred years or more has scarcely brought society any closer to the realization of the project of modernity. The opposite is the case: as the process of modernization did not bring forth the expected results, the project of modernity is itself being eroded. The project that guided modernization in the past, therefore, no longer serves as a guide to the future.

As a result, today society faces high fragmentation of actors and world views, a phenomenon also called post-modernism. Not only has the number of social actors sharply increased, but in the absence of a common reference point, given the erosion of the project of modernity, the different social actors' world views are becoming increasingly acceptable. Indeed, any social actor's reference point now seems to be legitimate, a phenomenon that was accelerated by globalization. The overall picture now reflects incoherence and absence of direction; it is not clear where continuation of this process of post-modernization leads to, nor which actors are the most legitimate to define its orientation.

In addition, the global ecological crisis has reinforced and accelerated this process toward post-modernism, rather than reversed it or slowed it down. It has led to more fragmentation, further eroded collective projects, and contributed to the multiplication of social environmental actors. If one expected that the ecological crisis would help refocus the project of modernity and give new coherence to the scattered actors and their world views, the opposite seems to have taken place: ecological considerations and actors became absorbed by the cultural trend toward post-modernism and helped push it along. As a result, the newly emerged environmental actors as well as the solutions they propose are equally scattered, fragmented, and incoherent.

Post-modernism has also translated into politics, and post-modern politics can be considered the new expression of societal fragmentation and the erosion of the project of modernity. This is problematic, as traditional politics is part and parcel of modernization, supposedly contributing to the coming of age of the project of modernity. Key characteristics of post-modern politics are the erosion of the nation-state as the most legitimate unit of action and the subsequent emergence of other equally legitimate levels of political action, local, regional, and global. The multiplication of political action units, such as grass-roots organizations, public interest groups, or NGOs, is paralleled by the decline of traditional political parties. NGOs, for example, are simultaneously active at various levels, which further confuses the overall picture. The other face of the multiplication of political actors and political action levels is the corresponding emergence of multiple objectives. These multiple objectives are not necessarily antagonistic; even more problematically, they are generally incomparable and, therefore, usually cannot be compromised. Substantive political objectives, generally all related to the project of modernity, such as equity, justice, and human rights, are increasingly replaced by expressive objectives, that is, basically, the call of various actors for the right to express themselves. Not surprisingly, post-modern politics has, therefore, mainly become a struggle for public attention. Political marketing and organizational efficiency, rather than content, have become the means to achieve it.

Environmental NGOs can, at least partly, be perceived as the expression of such post-modern politics: they emerge within the context of eroding traditional politics and the corresponding fragmentation and erosion of the project of modernity. Moreover, they are an expression of this very fragmentation. As such, environmental NGOs, like all other newly emerging actors, do contribute to the further post-modernization of society, that is, the fragmentation and erosion of collective goals. But, in today's global environmental crisis NGOs still have another function. To understand it, I must briefly recall the nature of today's crisis.

Today's global crisis: a vicious circle

If the process of modernization has been fuelled by a rational world view, mechanistic science, territorial expansion and conquest, nation-state politics, fossil fuel-based economic development, and scientific management, its most important driving force, it seems to me, is the increasingly artificial, yet heavily institutionalized separation between culture and nature, between society and the biophysical world. By separating the natural from the social sciences society has, indeed, set up a process of mutual reinforcement: on the one hand, some natural sciences and corresponding technologies contribute to the increased mastery over nature, thus producing the ingredients necessary for socio-economic development and modernization. Scientifically managing society's development on the other hand – for example through politics and education – guarantees that human, financial, political, and other resources are made available for the pursuit of mastering nature. Let me call this the 'development spiral'. Although this process had already led to unpleasant side-effects such as pollution and resources depletion in the 1960s and 1970s, it was believed at that time that the solutions to these problems could be found within the development spiral itself.

But in the 1980s – in the age of global ecology – this development spiral came to be questioned in a much profounder way. A new type of science, based on a new global awareness – the science of global bio-geo-chemical cycles – demonstrated that global limits have been reached, if not breached: the amount of carbon dioxide already present in the atmosphere, for example, will be sufficient to negatively affect the biosphere; so will the present state of depletion of the ozone layer. As a result, more and more people question whether the global environmental crisis can actually be managed along the line of the development spiral and the corresponding problem-solving strategies, such as more science and technology, better nation-state politics, more efficient economic growth and better education.

This becomes even more obvious if one looks at negative feedback loops: increased mastery over nature today negatively affects, via global environmental degradation, all societies. Such degradation will come on top of, reinforce, and further accelerate already existing societal trends which degrade the environment, such as the above-mentioned fragmentation and post-modernization. By doing so, environmental degradation will rapidly diminish society's options in effectively dealing with the crisis. In short, global environmental degradation and destructive societal consequences will reinforce each other in an ever-accelerating vicious circle. Let me highlight four such vicious circles:

First, environmental degradation is likely to put additional stress on society. Such stress will take the forms of social unrest and growing protest. More generally it will lead to increasing political instability. This might lead either to societal breakdown, or to increased social control. Both scenarios will make it more difficult to deal with the environmental causes of the societal crisis.

Second, coupled with population growth or other *a priori* unrelated problems, environmental degradation will exacerbate hunger and poverty, therefore accelerate migrations from the country to the megapoles, as well as from the less to the more developed countries. The urgency of these trends will force society to immediate action, thus diminishing society's options when it comes to effectively addressing environmental decline.

Third, environmental degradation is likely to create additional social and political conflicts, for example between winners and losers. Even if 'winning' might simply mean losing less than others, the cleavages so created between winners and losers will make concerted action more difficult. And, given the limited or even decreasing worldwide resource base resulting from, among others, environmental decline, conflicts between the North and the South, among specific nation-states, as well as within certain countries, are highly probable outcomes. Such conflicts will absorb resources and energies much needed for dealing with the global environmental crisis.

Finally, environmental degradation is likely to have negative effects on the psyche and the culture of all the planet's inhabitants. Rapid changes in the physical and in the social environment will almost certainly lead to further loss of roots and the erosion of cultural identity. More generally, the rapid changes in the physical and economic environment, coupled with the social transformations outlined above, will increase individuals' feelings of fear, anxiety and insecurity, with as yet unpredictable consequences. Such psycho-cultural changes will make the collective actions required to address the global environmental crisis increasingly unlikely.

In short, the global environmental crisis will exacerbate and accelerate existing destructive trends in society, which in turn will further degrade the biosphere. Or in other words, the development spiral turns, in the age of globally imposed limits to growth, into a 'vicious circle'. Social environmental learning is therefore neither about managing the environmental, nor the societal consequences of this vicious circle. Rather, it is about how to break out of it. Let me call this 'learning our way out'.[29]

Learning our way out

Given the existence of this vicious circle, traditional problem-solving approaches appear today, at best, to be inadequate. At worst, they are now counter-productive, tending to further accelerate the overall trend in environmental degradation and the simultaneous erosion of the socio-cultural basis for dealing with the global crisis. For example, public information campaigns conducted in this era of the atomized individual and already high environmental awareness are likely to result in apathy, cynicism and even despair. Solutions of a purely scientific or technological nature, especially if lacking in any social perspective, will further erode the very social and cultural resources which could have transformed them into meaningful social and cultural action. Traditional politics and policies, generally aimed at the promotion of economic growth, will become increasingly defensive and reactive: indeed, the traditional political approach at the nation-state level will be to save what is left of industrial development without offering any way out. And it is not just industrial development which must be transformed: radical changes in the very nature of economic development are imperative if environmental degradation and cultural erosion shall be halted.

In other words, the pursuit of exclusively economic, political, techno-logical or educational solutions will not be sufficient to solve today's increasingly global environmental crisis. Indeed, in the age of globally imposed limits to growth, it is very likely that each of these solutions stemming from the era of the development spiral will prove counter-productive. Only a change in perspective can help us learn our way through the crisis. Rather than trusting in the 'miracle' solutions mentioned above, we have to recognize the need for collectively learning our way out. All those actively promoting traditional problem-solving strategies must engage in this learning process. Experts promulgating counter-productive solutions should join groups of learners working collectively with real people on concrete problems. Teaching and preaching ready-made solutions to individuals must be replaced by collective, vertical, horizontal and cross-disciplinary learning. Such learning must be recognized as probably the only 'resource' still available to get us through and out of the ever-accelerating vicious circle.

Learning our way out will have to be a collective endeavour. There is no individual way out. Society must have to promote collective learning units which function within concrete biophysical limits. These limits, in turn, will have to feed back into the learning process. Therefore, learning our way out is not only about how to break out of this vicious circle, it is also about how to live sustainably within these limits, keeping in mind, in particular, that the vicious circle has already set into motion a process that will further restrict, not expand these limits. Globally, this means that learning our way out occurs against the background of a finite planet and a blocked horizon; locally, it will have to take place against the limits of local livelihoods, and natural, societal, and cultural constraints imposed upon them. Already we can conclude that some are more appropriate learning units than others. Nation-states and other societal actors whose only mission is development are probably not appropriate learning units. Villages, communities, cities, and some institutions might be more appropriate to start learning our way out.

As I suggest that environmental NGOs free themselves from traditional politics, change the reference point and privileged means of action, grow in numbers and interconnectedness, and become increasingly transnational, they contribute to societal change and transformation in yet another way: they become agents of social learning and therefore significant contributors to learning our way out. Indeed, rather than focusing on traditional politics, how to influence it and how to mobilize for it, environmental NGOs build communities, set examples, and increasingly substitute for traditional political action. They become agents of social learning, whereas social movements were actors of political change only. Taking traditional politics as the cultural model prevented social learning from taking place. Yet, their active role in fostering social learning is probably the most characteristic feature of environmental NGOs today – a feature social movement and third system theory can scarcely account for.

NOTES

1 See Ronnie D. Lipschutz, 'Heteronomia: The Emergence of Global Civil Society' (Paper presented at the annual meeting of the International Studies Association, Atlanta, GA, March 31 to April 4, 1992).
2 James Rosenau, *Turbulence in World Politics: A Theory of Change and Continuity* (Princeton, NJ: Princeton University Press, 1990).
3 Charles Hermann, 'Discussion of "Turbulence in World Politics" by James Rosenau' *American Political Science Review* 85 (1991) 1081–3, quote on 1083.
4 See Ted R. Gurr, *Why Men Rebel* (Princeton, NJ: Princeton University Press, 1970).
5 See Mancur Olson, *The Logic of Collective Action: Public Goods and the Theory of Groups* (Cambridge, MA: Harvard University Press, 1965); A.O. Hirschman, *Exit, Voice, and Loyalty: Responses to Decline in Firms, Organizations, and States* (Cambridge, MA: Harvard University Press, 1970).
6 See Bert Klandermans and Sidney Tarrow, 'Mobilization into Social Movements: Synthesizing European and American Approaches', in B. Klandermans *et al.*, eds., *International Social Movement Research*, vol.1 (Greenwich, CT: JAI Press, 1988).
7 See Herbert Kitschelt, 'Political Opportunity Structures and Political Protest: Anti-Nuclear Movements in Four Democracies', *British Journal of Political Science* 16 (1986) 57–85; Ferdinand Müller-Rommel, ed., *New Politics in Western Europe: The Rise and Success of Green Parties and Alternative Lists* (Boulder, CO: Westview, 1989).
8 See Alberto Melucci, *Nomads of the Present: Social Movements and Individual Needs in Contemporary Society* (London: Hutchinson, 1989).
9 See Joachim Raschke, *Soziale Bewegungen* (Frankfurt: Campus, 1985).
10 See Hanspeter Kriesi, 'New Social Movements and the New Class in the Netherlands', *American Journal of Sociology*, 94 (1989) 1078–116.
11 Alain Touraine, *The Self-Production of Society* (Chicago IL: University of Chicago Press, 1978); *idem, The Post Industrial Society* (New York: Random House, 1981); *idem, The Voice and the Eye: An Analysis of Social Movements* (Cambridge: Cambridge University Press, 1981); *idem, The Anti-Nuclear Protest* (New York: Cambridge University Press, 1983); *idem,* 'An Introduction to the Study of Social Movements' *Social Research* 52 (4) (1985) 749–87; *idem, The Return of the Actor* (Minneapolis, MN: University of Minnesota Press, 1988).
12 See Lynton Caldwell, *International Environmental Policy: Emergence and Dimensions* (Durham, NC: Duke University Press, 1990).
13 Claus Offe, *Contradictions of the Welfare State* (Cambridge, MA: MIT Press, 1984); *idem,* 'New Social Movements: Challenging the Boundaries of Institutional Politics' *Social Research*, 52 (4) (1985): 817–68.

14 Jürgen Habermas, 'New Social Movements.' *Telos*, 49 (1981): 33–7.
15 See Sidney Tarrow, *Social Movements, Resource Mobilization and Reform during Cycles of Protest: A Bibliographical and Critical Essay* (Ithaca, NY: Center for International Studies, Occasional paper no. 15, Cornell University, 1982); Mayer N. Zald and John D. McCarthy, eds, *The Dynamics of Social Movements. Resource Mobilization, Social Control and Tactics* (Cambridge, MA: Winthrop Publishers, 1979); Mayer N. Zald and John D. McCarthy, *Social Movements in an Organizational Society* (New Brunswick: Transaction Books, 1987).
16 See Kitschelt, 'Political Opportunity Structures'.
17 Marc Nerfin, 'Neither Prince nor Merchant – An Introduction to the Third System', IFDA dossier, no. 56 (1986): 3–29.
18 Ibid., 4.
19 Ibid., 6.
20 Ibid., 7–8.
21 David Korten, *Getting to the 21st Century: Voluntary Action and the Global Agenda* (West Hartford, CT: Kumarian Press, 1990), 185–92.
22 Nerfin, 'Neither Prince nor Merchant', 4.
23 Ibid., 5.
24 Ibid., 11–12.
25 Ibid., 4.
26 Korten, *Getting to the 21st Century*, 218–19.
27 Nerfin, 'Neither Prince nor Merchant', 16.
28 Ibid., 19.
29 See Lester Milbrath, *Envisioning a Sustainable Society: Learning Our Way Out* (Albany, NY: SUNY Press, 1989).

Part II
NGO relations

4 Advocacy and diplomacy: NGOs and the Great Lakes Water Quality Agreement

Jack P. Manno

On September 24, 1987, John Jackson, then vice-president of Great Lakes United (GLU), received a letter from the Canadian Minister of State for External Affairs, the Right Honourable Joe Clark. The letter responded to a series of letters GLU had sent to Environment Canada and External Affairs asking the Canadian government to include representatives of environmental interest groups in the delegation to bilateral talks with the United States over proposed revisions to the Canada–United States Great Lakes Water Quality Agreement (GLWQA). Similar requests had been made by Great Lakes United members in the United States to appropriate authorities there. Joe Clark's letter read, in part:

> With respect to your request for observer status at the bilateral review, you will appreciate that the presence of a binational nongovernmental group at the formal review of an international agreement by its signatories raises some interesting issues of propriety and precedent. Nonetheless, in view of Great Lakes United's credentials as a serious and responsible group and our collective interest in ensuring the best possible review of the Agreement, I am pleased to invite you and one other member of the Canadian section of Great Lakes United to participate as observers to the Canadian delegation.

This case history explores some of the issues of propriety and precedent referred to by the Minister of External Affairs. These issues not only bear on Canada–US relations but also reflect similar issues raised elsewhere in international environmental relations. Non-governmental organizations (NGOs) are increasingly insisting on the importance and value of their participation. This inquiry into the participation of a non-governmental organization in bilateral Great Lakes negotiations creates an opportunity to examine empirically the development of NGO strategies, cross-sectoral dynamics, internal organizational development and the relationships between institutions and ecosystems. The inquiry draws heavily on the observations and personal records of the participants to the negotiations. From these and the historical record of Canada–US efforts to jointly manage and protect the Great Lakes, lessons are drawn that may be useful for understanding other cases in which NGOs play a critical role in international environmental relations.

NGOs may be remapping the terrain of international environmental affairs, but studies of international environmental relations are still mostly presented from the perspective of national governments and through the academic lenses of international studies. This study, by contrast, is a narrative and interpretive history of the role played by NGOs in the events leading to the adoption of the 1987 protocols to the Great Lakes Water Quality Agreement. The case demonstrates that one cannot hope to understand US and Canadian environmental relations without considering the policies, strategies, and actions of the NGOs. Furthermore, the NGOs themselves cannot be understood without placing their organizational development and the evolution of their influence within the context of binational relations and regional politics. Lastly, and even more importantly, neither the NGOs nor the nation-states can be understood apart from the geographical realities and the changing ecological characteristics of the Great Lakes Basin ecosystem itself, which ultimately shape the region's economies, demarcate its political boundaries, and affect all enterprises within its realm.

The people of Canada and the United States share the world's most extensive boundary waters, made up for the most part by the Great Lakes-St Lawrence River, the world's largest system of fresh surface water, draining nearly 200,000 square miles of land. For several centuries the Great Lakes region was a powder keg of tensions as French and British armies, American and Canadian settlers, and the indigenous nations competed for navigational access to the continent's interior and for control over its abundant fur-bearing animals and other sources of wealth. By the twentieth century, political powers in the Great Lakes region had concentrated in the British Commonwealth government of Canada and the federal government of the United States.[1] The two states began to focus on cooperation, first to recognize each other's rights to peaceful navigation and later to respond to what was becoming a large-scale pollution catastrophe.[2] Since 1972 the Canada–US GLWQA has served as the reference point for cooperative action to reverse trends of deteriorating water quality. The GLWQA is, according to the International Joint Commission, a 'milestone document, one of the first international statements that technical, diplomatic and administrative approaches to resource management need to be considered in terms of holistic ecological concepts.'[3]

These holistic ecological concepts are manifested in the US–Canada GLWQA in the following ways:

1 acceptance of a definition of the Great Lakes Basin Ecosystem that includes human beings and the adoption of the concept of 'ecosystem integrity' as the goal for environmental restoration;
2 reliance on planning and government intervention on the scale of ecosystems across arbitrary jurisdictional boundaries;

3 recognition that biological and ecological processes interact with physical and chemical ones to bioconcentrate particular classes of persistent toxic compounds, defined as critical pollutants, that require extraordinary regulation;

4 recognition that land-use practices in one part of the basin could significantly affect ecosystem quality in downstream, distant parts of the basin.[4]

Forces driving change in Great Lakes institutional arrangements include biophysical alterations of the ecosystem, improvements in scientific understanding of ecology, toxicology, limnology and other relevant sciences, changing political realities and the evolution of concepts and laws concerning government responsibility for the health of ecosystems and public participation in decisions. These forces, both environmental and social, are expressed in changes in the institutional structures of governance. The process of change through experimentation and response is sometimes referred to as 'social learning'. The evolving Great Lakes governance structure is one example of the multi-faceted partnerships being experimented with throughout the world.[5] Indeed, the Canada–US GLWQA with its espousal of an 'ecosystem approach' to environmental protection has been promoted as a model for global institutional arrangements.[6] The Great Lakes experience may indeed be suggestive and lessons drawn here may fruitfully be applied to other shared ecosystems, including the biosphere as a whole.

Issues concerning management of a shared ecosystem have at times seemed to dominate Canada–US relations. The complex of organizations and individuals involved in Great Lakes water quality activities forms an evolving governance structure[7] comprising bilateral institutions, federal, state and provincial agencies, the 'expert community'[8] of professional and informal networks of scientists, as well as environmental advocates, native activists, financial, industrial and tourism interests, hunters and anglers, the press and others. Within this governance structure private non-governmental organizations play a major role.

In examining the international relevance of the Canada–US Great Lakes relationship the growing influence of non-governmental organizations in both domestic and binational Great Lakes policy-making stands out. Environmental NGOs have played an important role, particularly in the 1980s, in defining the issues in each country and determining the bilateral institutional responses to those issues.

In 1989 the IJC wrote in its Fifth Biennial Report on Great Lakes Water Quality:

> The emergence of strong, sophisticated and effective non-governmental organizations over the past decade has been a positive development. Composed of many thousands of Great Lakes basin residents and others from both sides of the international boundary, these organizations are important in focusing political attention on the integration of Agreement objectives into domestic priorities and programs. They are instrumental in

encouraging governments to provide the resources necessary to implement the Agreement and actively promoting environmentally conscious
behavior among their own membership and the public at large. As such
these organizations fill a distinct niche in the Great Lakes institutional
framework.[9]

In this study I focus on the formal bilateral review of the GLWQA and the
negotiations leading to the 1987 amendments. Perhaps one of the most
significant aspects of the negotiations does not actually appear in the
document which the parties signed. It is, rather, the manner in which the
review and amendment negotiations were carried out. For the first time in the
long history of formal Great Lakes negotiations, representatives of three non-
governmental organizations – Great Lakes United (a binational coalition),
Sierra Club, and National Wildlife Federation – were invited by the State
Department as observers to participate as members of the US delegation.
Likewise, the Ministry of External Affairs invited two representatives of
Great Lakes United to serve as observers in the Canadian delegation.

In this study a brief recounting of the history of US–Canadian affairs as
they pertain to the boundary waters and the GLWQA forms the basis of an
analysis of the biophysical, social, and political factors that underlie this
governance structure. I describe the history and development of one of the
three NGOs, Great Lakes United, which played the key role in achieving
observer status at the negotiations as a prelude to a recounting of the 1987
negotiations. The study concludes with lessons drawn from this case that bear
on further study of international environmental negotiations and NGO and
government strategies.

The GLU representatives and the other observers did far more than
observe. They were thoroughly involved in discussing every aspect of the
agreement and bringing with them a high degree of technical knowledge and
an ability to articulate technically supported positions. The NGO observers
had the advantage of being part of a binational network of advocates. They
were thoroughly familiar with the proposals from both parties and the internal
politics of each and, therefore, had a deep understanding of the various
proposals. In the end, the new annexes added to the GLWQA were significantly shaped in both wording and intent by the persuasive efforts of the NGO
'observers'. Their efforts gave political expression to several long-standing
recommendations that had arisen from several International Joint Commission boards and other forums, such as the Anticipatory Planning Workshop,
the Pollution from Land-use Activities Reference Group (PLUARG) and the
Hiram Workshop on implementing the ecosystem approach, held during the
1970s.[10] For example, the NGOs placed on the agenda and won requirements
for public participation in GLWQA implementation, in particular in the
remedial plans required by Annex 2. They argued for and won stricter and
narrower definitions of 'point source impact zones', in Article IV, insisting
that no exceptions for industrialized embayments be made to the parties'

commitment to the virtual elimination of persistent toxic substances throughout the Great Lakes ecosystem. NGO representatives also successfully
supported a redefinition of critical pollutants and the elimination of gender-
specific language from the GLWQA. In addition to these changes the range
of subjects covered under the amended Agreement's provisions was expanded, partly as a result of the NGO's efforts, to include airborne pollutants,
pollution from agricultural and land-use activities, contaminated groundwater and wetland protection. Perhaps most importantly, the presence of
the NGO delegation helped prevent the possibility of political mischief in
the form of last-minute alterations to the agreement text, emanating from
ideologues in conservative governments in Canada and the United States.

Finally, one additional result of the 1987 agreement apparently had not
been anticipated fully by the negotiators, including the NGOs – the weakening of the International Joint Commission as an international institution. The
NGOs seem to have had little appreciation of the role the International Joint
Commission (IJC) and its working boards as forces for moral suasion. When
the NGOs considered the IJC at all, during the process leading to the 1987
protocols, it was mostly to criticize the Commission's lack of implementation
authority. The negotiators accepted the recommendations of several observers
that government accountability be built into the GLWQA. As a result, the
new annexes clearly charge the parties – the US and Canadian governments –
with the responsibility for implementation and reporting on progress. The
resulting agreement led to a new binational committee structure that dupli-
cated the existing set of IJC boards and committees. This, coupled with
criticisms by the NGOs of the government members of the IJC boards as
being compromised by conflicts of interests, led the IJC commissioners in
1991 to dissolve the committee structure, effectively terminating an important intergovernmental forum.[11]

BACKGROUND OF CANADA–US GREAT LAKES WATER QUALITY AGREEMENT

The negotiating history leading to the GLWQA dates to the late nineteenth century, when significant advances were made in waterworks engineering and economic development. Along with advances in technology came
plans for constructing major works with the potential for altering parts
of the Great Lakes hydrological system. Proposed canals and dams raised
concerns about water resource rights. Potential and actual disputes over
such rights recurred often and were handled through a cumbersome series
of diplomatic exchanges between Dominion authorities in the British government and the US State Department.[12] Because of the lengthy diplomatic
correspondence between London and Washington, minor disputes frequently festered. A proposal for a Chicago drainage canal to divert Lake
Michigan water into the Mississippi River basin, and another for a dam at
the outlet of Lake Erie, were two of the most controversial. Both were

initiated on the US side, with little consideration given to the possible impact on Canadian rights and resources.

Dominion representatives pressed for a treaty that would protect Canadian interests, which they felt were constantly being pitted against US economic might. Canada sought a strong treaty enforced by a commission with wide-ranging authority. The United States, however, preferred measures that would not impinge on national sovereignty rights. The Boundary Waters Treaty of 1909 was the compromise result. It established a body, the International Joint Commission (IJC), empowered to act only upon those cases jointly referred to it by the parties. It held no authoritative powers over the two participating states to ensure compliance with its recommendations.[13] Still, its structure did offer a unique approach to international problem solving. The six commissioners, three Canadian and three American, were expected to represent the commission, not their home countries. Decisions were to be made by consensus and, to insulate commissioners from political pressure, no record was kept of the decision-making process itself.

The failings of the IJC have not been caused by disputes between the parties, because almost every decision has indeed been made by consensus. Rather, shortcomings have resulted from the complex and difficult problems of the Great Lakes ecosystem itself and from the limited powers and resources the commission has had to provide solutions and gain cooperation from the parties. The evolution of these issues and institutional arrangements are central to understanding the significance of the GLWQA protocol negotiations in 1987 and the precedent set by the involvement of environmental NGOs.

GREAT LAKES WATER POLLUTION: A CATALYST FOR CHANGE

Interest in water pollution antedated the 1909 Boundary Waters Treaty. At that time, typhoid fever was a major health problem in the United States and Canada, and a clear link had been established between polluted water and the spread of typhoid.[14] Investigative studies of Lake Michigan and Lake Erie suggested the need for federal public health legislation. As a result, the Boundary Waters Treaty addressed water pollution in Article IV: 'It is hereby agreed that the waters herein defined as boundary waters and waters flowing across the boundary shall not be polluted on either side to the injury of health or property on the other.'

This article has grown immensely in importance since 1909. It has provided the basis for IJC investigations into water pollution and water quality issues, and eventually provided the rationale for the GLWQA.[15] The IJC received its first reference to investigate water pollution in 1913. Following investigative studies in the connecting channels, both the US and Canadian commissioners issued preliminary reports that were dramatic in their urgency. The language expressed deep concern: 'The situation along the frontier which is generally chaotic, everywhere perilous and in some cases disgraceful [and the conditions] imperil the health and welfare of the citizens in substantial contravention of the spirit of the Treaty.'[16]

The commission's 1918 reference report cited sewage from vessels, cities, and industries as major causes of the pollution problem. To address the pollution problem the commission requested that it be given sweeping powers to regulate and prohibit sewage pollution. The government response was a request to the commission that it draft a water quality treaty. By 1920, however, widespread acceptance of water filtration and chlorination had effectively eliminated typhoid fever; the urgency of eight years earlier had dissipated and the momentum for a water pollution treaty was lost.

The spread of typhoid fever had been a dramatic, high-profile water pollution crisis. The adoption of widespread public health measures in cities around the basin effectively removed water quality issues from the binational agenda for the next two decades. But the processes of ecosystem degradation continued, despite progress in protecting humans from waterborne diseases.

From the time of European settlement, human-induced stress of the Great Lakes accelerated, to the verge of ecosystem crisis. Logging throughout the basin raised water temperatures and choked the tributaries with the silt of eroded riverbanks. When streams were dammed for mills, salmon lost access to spawning grounds and habitat. Unrestrained fishing drove the populations of top predator species to unsustainable levels. The Welland Canal opened the upper lakes to access by sea lamprey from Lake Ontario. Seagoing vessels brought in a myriad of other organisms. The cold water fishery was further devastated by oxygen depletion brought on by algal blooms stimulated by sewage and other inadvertent forms of fertilizer.[17]

By the time the general public took serious notice in the 1960s the momentum of large system modification had already caused considerable damage. Scientific concern for the health of the lakes and public demand for action led the governments of Canada and the United States to ask the IJC in 1964 for a study of water pollution problems in the lower lakes, Lakes Erie and Ontario, and the St Lawrence. The study took six years to complete, but the IJC's response of 1970 called for an international clean-up effort, urging the governments to develop programmes to reduce phosphorous inputs and to agree on controls and/or regulations on several pollution sources. Those six years also saw a dramatic outpouring of public concern about the environment.[18] Negotiations leading to the Great Lakes Water Quality Agreement of 1972 began almost immediately after the governments received the report. By that time major fish die-offs, beach closings, mounds of rotting seaweed, and river surfaces that actually had caught fire had had their effects. The visible outcome of sewage and fertilizer pollution and the resulting eutrophication of the lakes served as the motivating backdrop for the GLWQA negotiations, and the control of phosphorous inputs was its primary remedial strategy.

THE 1972 GREAT LAKES WATER QUALITY AGREEMENT

In the 1972 GLWQA the parties expressed their determination to 'restore and maintain the chemical, physical and biological integrity of the Great Lakes'. The agreement also gave the IJC additional responsibilities for:

1 collecting, analysing, and disseminating information on the operations and effectiveness of government programmes to improve water quality of the Great Lakes;
2 tendering advice and recommendations to federal, state, or provincial governments for dealing with water quality problems;
3 assisting in the coordination of joint efforts to control pollution, including the discharge of phosphorus into the lakes.

These new powers, in effect, constituted a permanent reference. The commission was no longer required to wait for the parties to refer specific questions to it before commenting, criticizing, and offering advice. To carry out its new functions under the GLWQA, two new binational IJC boards were established: the Water Quality Board and the Great Lakes Science Advisory Board. The Water Quality Board serves as the principal advisor to the IJC on all matters pertaining to the GLWQA.[19] The Science Advisory Board serves a broader, less-focused purpose, advising the commissioners on research and scientific matters and calling attention to new and emerging issues.[20]

The new boards made available to the commission a source of technical and managerial expertise, allowing the commissioners to comment broadly in the biennial reports they issued under the GLWQA. The boards' research and reports did several things besides informing the commission. They clarified and documented the causes of water pollution, recommended government action, and alerted the public. The boards also stimulated and became part of a new complex of working relationships among US and Canadian natural scientists, ecologists, bureaucrats, and policy scholars with links to both governments and the new environmental NGOs of the 1970s.[21] The seeds of this new 'expert community' lay in earlier collaborative efforts, such as the Northington study of Lake Erie begun in 1960, work done under the 1964 US Water Resources Research Act, preparation for the 1972 International Field Year on the Great Lakes (IFYGL), and the Canada-United States Interuniversity Seminar (CUSIS) involving faculty members of twenty Canadian and US colleges and universities in the early 1970s.[22]

The agreement's remedial strategies grew principally out of the recommendations of the two reference groups constituted in 1964 to study the lower Great Lakes and the St Lawrence River where the pollution was most conspicuous. To expand on the previously completed studies the GLWQA called for two major follow-up studies: one on the upper lakes and the other on the diffuse sources known as 'nonpoint pollution'. Two IJC study groups were formed: the Upper Lakes Reference Group and the Pollution from Land-Use Activities Reference Group (PLUARG). The Upper Lakes Reference

Group played a key role in the evolution of public participation in IJC reference studies.[23] The group set up to carry out the studies in support of the Upper Lakes reference decided to hold a series of public workshops to explain the issues and solicit opinions. The group contracted with Great Lakes Tomorrow, the first binational Great Lakes citizens group that had been set up to educate the public and facilitate public involvement in Great Lakes decisions. This experience provided the basis for future citizen involvement in IJC activities.[24]

The advances the Upper Lakes Reference Group made in public participation were taken to new levels in the massive ecological study known as the Pollution from Land-Use Activities Reference. PLUARG consisted of more than one hundred investigators in a five-year study of pollution from agriculture, forestry, and other land uses. As the first IJC reference dealing with the entire Great Lakes basin and involving public consultation panels from throughout the basin,[25] it proved to be very important not only in expanding scientific understanding of multiple sources of pollution but also in laying the groundwork for an ecosystemic approach and expanding public participation in IJC activities.

Although there was little precedent for involving others besides government-appointed experts in IJC investigations, the logic of public participation in PLUARG was relatively simple. The reference group was being asked to study an impossibly large subject across a vast geographic area: the set of activities within the Great Lakes drainage basin – agriculture, suburban development, highway construction, and so forth – all of which either added polluting substances to the ground, ultimately to reach the lakes, or increased erosion, and subsequently, the run-off of silt and soil into the lakes. If such activities were to be controlled, they would ultimately be controlled at the local and even individual level. For PLUARG to derive recommendations based on the actual pattern of life activities in the Great Lakes basin, and for those recommendations to have any chance of successful implementation, the cooperation and support of large numbers of politically influential individuals would be required. Public participation in PLUARG was premised on a general trend toward democratization of the decision-making processes usually left to experts,[26] on the opinion of the experts that the public had valuable, personally obtained information to share and that the public would need to be mobilized before PLUARG could achieve its ends. The stated objectives of the PLUARG consultation panels were to gain public support of the final PLUARG report to the IJC and to lend credibility to both PLUARG and the IJC.[27]

The reference group organized seventeen citizen panels around the basin, nine in the United States and eight in Canada, made up of several hundred citizens.[28] The consultation process was unique in being characterized by its geographic extent, binational involvement, and use of citizen panels.[29] Citizens advised PLUARG on all aspects of the study. Their involvement not only had a direct impact on the final report but also positively influenced people's attitudes toward the GLWQA. It was successful in gaining both support and credibility as hoped.[30] The *Great Lakes Communicator*, a publication of a state-federal water resource planning agency, reported that 'public involvement in PLUARG had been a useful and successful aspect of the study indicating that public involvement should continue to be a part of future management strategy'.[31] Although PLUARG panel reports and the final report to the IJC recommended expanded public education and participation, no provisions were made by the IJC or the parties for the continued involvement of consultation panel representatives in implementation of PLUARG's recommendations.[32] Despite this, Mimi Becker, who with Sally Leppard ran public workshops to train interested citizens for participation in IJC hearings, maintains that along with the work on the Upper Lakes Reference Group the PLUARG efforts

> set the precedent for opening up the IJC annual meetings so that citizens could have more than just the privilege of asking questions during the press conference, and provided the basis for the IJC to deal more substantively with informed members of the public.[33]

In addition to opening up the process to the public in unprecedented ways, the research accomplished under these new investigative initiatives furthered the ecological understanding of the Great Lakes and provided a scientific base of information that served as the impetus for the 1978 GLWQA. Studies confirmed the impacts of cross-media pollution, such as acid rain and nonpoint source pollution from agricultural lands and groundwater sources, and, thus, substantiated the need to consider more than just water quality in efforts to curb pollution.[34]

The 1972 GLWQA was in force for five years, after which it was to be revisited by the parties. In the years between 1972 and 1978 progress was made in reducing phosphorous inputs, through sewage treatment and the gradual elimination of phosphorus from laundry detergents.[35] The eutrophication problem was on its way to being resolved. With this success, the problem of toxic industrial chemicals and pesticides present in the flesh of fish and other animals, previously masked by the more visible problems of eutrophication, re-emerged as the focus of concern in the Great Lakes.

As early as 1963, studies of herring gull eggs in Lake Michigan concluded that thinning shells and poor reproductive success was probably associated with concentrations of DDT and its toxic metabolite, DDE, that the birds received from their diets of Lake Michigan fish. In 1968 mercury from the chlor-alkali wastes being dumped into the lakes and tributaries was measured in the sediment and fish of Lake Ontario. In 1971, common terns in Hamilton Harbor were discovered with deformed cross bills, an apparent result of the chemical stew of PCB, DDT, and hexachlorobenzene found in the eggs. Mirex, an organic chemical fire retardant and pesticide, was discovered in fish in the early 1970s.[36] By the mid-1970s, states and provinces were routinely issuing warnings about eating the fish from the lakes, and several commercial fisheries were closed.

The chemicals of primary concern are synthetic organic chemicals produced directly or as by-products of industrial processes. Sources include industrial and municipal outfalls; contaminated air and rain; leaking landfills; previously contaminated sediment resuspended by currents, dredging, and storms; agricultural practices; and the widespread household use of solvents and pesticides. They represent a source of biochemical stress new to the industrial era that Great Lakes creatures had never encountered, and for which they had evolved few mechanisms to cope. The most serious threat came from chemicals that did not break down through metabolic action and those that were insoluble in water and concentrated in fat. Their resulting environmental persistence means they circulate and recirculate unchanged through the ecosystem's physical and biological pathways, gradually becoming ubiquitous throughout the system. Because they are stored in fatty tissues and accumulate, they concentrate as they rise up the stages of the food chain. For instance, PCBs are bioaccumulated 25,000,000-fold in Great Lakes food webs from water to bald eagles' eggs. Hence, minute amounts of certain chemicals can become large problems throughout the whole system.[37]

The toxics problem was significantly more complicated than the primary problem addressed by the 1972 agreement – nutrient pollution – which could be traced comparatively easily to municipal sewage systems and phosphorus in detergents. The solutions to nutrient pollution – sewage treatment plants and detergent phosphorus bans – although expensive, were manageable with the participation and coordination of existing state and provincial governments. By contrast, the problem emphasized in the 1978 agreement, toxic contamination, could not be solved by a single jurisdiction nor without substantial changes in industry and consumer practices. The agreement needed, therefore, to break new ground in international cooperation and institutional arrangements.

The lesson taught by the presence of toxics in the Great Lakes was that society ignores the interrelationships of the natural system at its own peril. By the time levels of pollution reach the point where damage is apparent, governments face dwindling choices for correcting the problem. Clean-up costs are exorbitant and restoration may be impossible. The only pollution policy that makes sense is prevention – that is, understanding how stresses are likely to alter the ecosystem and to eliminate those stresses that are preventable and minimize those that are not. From such realizations came the case for policy based on ecosystem science and a subsequent commitment by Canada and the United States in the GLWQA to an ecosystem-based approach to restoring the integrity of the Great Lakes.

The revision of the GLWQA signed in 1978 greatly expanded the definition of the problem, as reflected in the agreement's area of purview. After recognizing that the problems of toxics in the Great Lakes water could not be resolved by actions focused on the lakes alone, the 1978 revisions extended the scope of the GLWQA to the entire Great Lakes ecosystem, including the land surrounding the lakes and the inflowing streams. In addition to extending the physical boundaries, they expanded the concept of water quality and acknowledged the interdependence of all components of the ecosystem, including humans.[38] The 1978 GLWQA defined the Great Lakes ecosystem as 'the interacting components of air, land, water and living organisms, including humans, within the drainage basin of the St Lawrence River' (Art. I, g).

The 1978 agreement expressed several additional concerns in response to the findings of PLUARG, nonpoint source pollution, and the effects of air pollution on water quality. The US and Canadian governments also agreed that

> the discharge of toxic substances in toxic amounts be prohibited and the discharge of any or all persistent toxic substances be virtually eliminated (Art II, a) that the philosophy adopted for control of inputs of persistent toxic substances shall be zero discharge (Annex 12).[39]

These two aspects of the GLWQA – the ecosystem approach to environmental protection and zero discharge of persistent toxics – derived from the growing awareness of ecology and the nature of the toxics problems.[40] The adoption of these concepts within a binational agreement is of major international importance. The challenge facing the governments in the region is how to translate an ecosystem approach and zero discharge into meaningful action feasible within the constraints presented by each nation's federal structures and political cultures.[41]

The International Joint Commission, in its *Second Biennial Report,* issued in December 1984, wrote:

> Existing resource management approaches which partition the environment into separate components of land, water and air with associated biota are recognized as inadequate since management of a resource component in isolation from adjacent or interacting components would likely produce short-sighted strategies to protect one component of the environment at the expense of another. Because existing environmental and resource programs are separated, compartmentalized and spread throughout various bureaus, agencies, ministries and departments, the new approach requiring a holistic overview entails, at the very least, a reorganization of thinking, and perhaps a reorganization of institutional arrangements.[42]

It may be evident that fundamental institutional change is necessary before an ecosystem approach to environmental protection becomes a reality, but institutional arrangements seldom reorganize themselves without pressure from outside forces. The participation of environmental NGOs in the decision-making process, insofar as it encourages governments to be accountable for their ecosystem commitments and brings new and creative ideas into the institutional dialogue, may be elements in closing the gap between ecosystem rhetoric and action.

Of all the problems with the lakes, the ubiquitous presence of industrial chemicals and pesticides that taint the lakes and compromise the health of its living creatures has been the one that has most taxed the creativity and resources of government environmental agencies. The seeming intractability of the toxics problem has brought into question the effectiveness of accepted regulatory policy and structures. As a result, political space to challenge government willingness and ability to protect the environment has been opened.

Several non-governmental organizations have stepped into that space, presenting alternative approaches to environmental protection. They have pushed their agenda on many levels: local, state and provincial, national and international. Claiming a stake in the entire ecosystem regardless of borders, they have acquired legitimacy as defenders of environmental interest. They have gained leverage against the parties and other actors by communicating and strategizing across national boundaries and by using the Great Lakes Water Quality Agreement and, in particular, the agreement's espousal of the ecosystem approach and its goal of zero discharge as their own. As the National Wildlife Federation's Tim Eder has said, 'It's always important to have goals against which to measure governments' progress, all the better if it's something the governments themselves have put out there.'[43]

The environmental NGOs in the Great Lakes region have often played this role *vis-à-vis* the GLWQA, pushing the institutions to find ways of implementing the various programmes outlined in the agreement.[44] The remainder of this chapter traces the evolution of the NGO role to the point where NGOs formally joined the two federal governments in amending the agreement in 1987.

The three NGOs invited as observers to amendment negotiations – Sierra Club, National Wildlife Federation (NWF) and Great Lakes United (GLU) – had each pressured governments in their own ways to implement the GLWQA. The Sierra Club, although originally a California association focused on the Sierra-Nevada Mountains, has, since 1945, grown into a nationwide organization with membership of nearly half a million. Its expanded purpose, according to its public literature, is

to explore, enjoy, and protect the wild places of the earth; to practice and promote the responsible use of the earth's ecosystems and resources; to educate and enlist humanity to protect and restore the quality of the natural and human environment; and to use all lawful means to carry out these objectives'.[45]

In 1986 the club increased its level of political activity in the Great Lakes by initiating the Great Lakes Federal Policy Project with funding from the George Gund and Joyce Foundations. Along with Great Lakes United the project coordinates an annual Great Lakes Washington Week, which brings activists to Washington to meet with Congressional representatives and EPA officials to gain hands-on experience with federal environmental policy-making and to raise Great Lakes issues at meetings and hearings. The project also publishes a monthly report whose goal is to 'provide timely information on federal actions affecting environmental quality of the Great Lakes' and to 'report on the activities of Congress, key agencies and other negotiations, covering issues from pollution control to appropriations.'[46] The project has offices in Washington, but is closely coordinated with Sierra Club's Midwest regional office and is led by the region's director, Jane Elder. Elder, along with GLU's Tim Eder and the National Wildlife Federation's Mark Van Putten, formed the NGO observer group on the US delegation to the 1987 agreement negotiations.

The National Wildlife Federation was founded in 1936 'to educate the public about conservation as well as the symptoms and the solutions to environmental abuse and neglect' (NWF brochure). The Great Lakes regional office in Ann Arbor, Michigan, has focused on the effects of toxic chemicals on fish and wildlife and on political and legal pressure to reduce the input of toxics to the lakes.

Great Lakes United, as a transnational coalition of organizations, including Sierra Club locals and National Wildlife Federation affiliates, was most involved in monitoring the GLWQA. The evolution of GLU's organizational structure, its positions and strategy, therefore, is presented in the following pages. GLU's history and its participation in the review and amendment process for the 1987 protocols to the GLWQA is highlighted primarily for the following reasons.

1 'As a coalition of sportsmen, environmental, conservation, labor, business, community organizations, and individuals from eight Great Lakes states and two Canadian provinces' (GLU promotional brochure), GLU represents many organizations in *both* nations. These member organizations have their own contradictory interests but have agreed to suspend those disagreements to cooperate for what they perceive as the benefit of the ecosystem. GLU encourages personal identification as 'citizens' of a watershed. This identification with ecozones, or bioregions, challenges presumptions of the predominant importance of national interests, presumptions that are, as noted later, already undermined by the nature of the environmental issues under discussion. This shift in presumptions allows consideration of questions of definitions and conflicting concepts of public, regional, national, and ecosystem interests.

2 As a coalition of organizations from throughout the basin, GLU includes groups with a broad spectrum of interests, from radical environmental activists to conservative national rifle association affiliates, from state governments and major academic research organizations to neighbourhood environmental clubs, from car workers to the Association of University Women, from organizations working for native sovereignty to sports clubs opposed to special treaty privileges. It has had to nurture carefully the shared assumption of mutual interests in the coalition, while speaking with

a clear and consistent voice on behalf of environmental protection. Only by representing a large and diverse constituency has GLU found a seat at the table of international negotiations, and only because it has had strong leadership able to act independently was it able to take the seat there.

3 As a binational organization, GLU has credibility when dealing with binational issues and, thus, has played a greater role in the Canada–US dialogue than advocacy groups operating in either nation exclusively.

4 Several commentators and scholars have remarked on GLU's effectiveness.[47] As noted later, GLU was particularly effective in developing and implementing a strategy for influencing the way the governments fulfilled their responsibility for reviewing the GLWQA in 1987.

5 The origins and history of GLU provide an example of the interactions between ecological and institutional factors, interest group dynamics, and nationalism. It is common to assume that examinations of international relations will draw upon information concerning the history and culture of the states involved. It is likewise useful to consider the unique history and culture of each NGO involved.

GREAT LAKES UNITED: BUILDING PUBLIC CONSENSUS AND THE POLITICAL WILL TO IMPLEMENT THE GREAT LAKES WATER QUALITY AGREEMENTS

Not surprisingly, the issues that have engaged concerned citizens in the Great Lakes have changed, along with the chemical, ecological, and social transformations described previously. Early in the twentieth century, public health reformers in cities across the region led the push for drinking water treatment and sanitation. The preservationist movement that gave rise to the Sierra Club and other groups in the United States[48] had an impact on the Great Lakes region, most notably in the effort to protect the Indiana Dunes from the industrial developments concentrating on the Southern shore of Lake Michigan.[49] The Canadian environmental movement has evolved from slightly different origins and influences, although toward similar goals of preservation and conservation. In the 1960s small environmental organizations came into being throughout the region. Most of them focused on specific evidence of pollution problems in their immediate area: fouled beaches in Erie, Pennsylvania; concerns about drinking water safety in Toronto; alewife die-offs in Lake Michigan; the decline of lake trout fishing in Irondequoit Bay; efforts to protect St Lawrence riverbanks and islands from the effects of seaway activities.[50] In many of these situations the public concerns over the obvious effects outpaced their knowledge of the causes of pollution. Governments were unable to respond to citizens' concerns with definitive answers.[51] This gap stirred many to turn to the new environmental organizations, which placed the blame squarely upon industry practices and government neglect. Heightened awareness of the environmental problems in the late 1970s, as well as growing environmental activism, resulted in a proliferation of new organizations in the Great Lakes basin and throughout the United States and Canada.[52] Concerns, particularly in the Lake Ontario region, reached new heights with the dramatic publicity surrounding the Love Canal contamination crisis. Organizations like Pollution Probe in Toronto pointed out that the same chemicals which were driving residents of Love Canal from their homes were leaking from scores of waste sites along the Niagara River. Toronto's drinking water intake pipes were only 50 km directly across the western basin from the mounth of the Niagara River where it drains into Lake Ontario.

Not only were environmental organizations in both Canada and the United States becoming increasingly involved in Great Lakes issues, they were also occasionally collaborating with each other across the border. Pollution Probe and a group called Operation Clean Niagara, based in Niagara-on-the-Lake, Ontario, received 'friend of the court' standing in lawsuits involving the dioxin-contaminated Hyde Park landfill in New York State, where leachate was trickling down the walls of the Niagara gorge into the river. Probe worked closely with a local coalition that had been heavily involved in Love Canal issues, the Ecumenical Task Force of the Niagara Frontier.

The exchanges between Canadian and American activists provided lessons for both, as each saw aspects in the other's legal institutions and administrative cultures[53] that they coveted. Canadian public servants in general have more discretionary authority to take action. As a result, when Canadian environmentalists participate in public consultation exercises, they are more confident that they are speaking with individuals who can make decisions. A tradition of public consultation has existed in Canada, and agencies often provide travel and other support to Canadian private groups to facilitate participation. As a result of this very independence, however, Canadian administrative decisions are less open to legal challenges in the courts, and Canadian activists, therefore, often look longingly across the border to a public armed with what looks from the Canadian perspective more like real power, in the form of access to environmental litigation.[54] Associated with these different traditions of public participation, Canadian and American groups also differ in their levels of political independence, which are the direct result of different tax laws. The Canadian tax system makes it much more difficult for organizations with any political aims to qualify for tax-exempt status. As a result, Canadian groups have less access to private and foundation funding, and often rely on the government for the bulk of their income. To Americans, this government support has often appeared to represent a more generous form of democracy, in that Canadians were, in some respects, being paid to challenge their governments.

Two issues surfaced in the late 1970s that fostered a sense of shared interest among environmentalists, government officials, and many businesses across the basin and in both countries: proposals to divert Great Lakes water to the drought-stricken Midwest of the continent, and reconsideration of the possibility of winter navigation on the Great Lakes–St Lawrence Seaway. In addition, many of the traditional sports and conservationist organizations began paying increased attention to threats to fist and game and their habitats from pollution and encroachment by human activities.

The revival of interest on the part of the shipping industry and the Army Corps of Engineers in the possibility of keeping the St Lawrence Seaway open throughout the winter months stirred considerable public protest. Normally, the seaway opens in early April and remains open into early December. During the winter months the seaway's customers switch to rail, trucks, and storage. Winter navigation would increase both the seaway's revenue season and its convenience and attractiveness to customers. Thus, over the years since the opening of the seaway in the 1950s, proposals have been regularly put forth to extend the season through the use of ice-breakers and underwater dams to keep locks and channels open. Just as regularly, environmentalists and riverside residents have raised concerns about accidental oil and chemical spills dispersing beneath the ice, bottom sediment and fish spawning areas being scoured by ice churned under by passing boats, shoreline erosion by tanker wakes and broken ice, and the disruption of a variety of sensitive winter fish and wildlife habitats.

Winter navigation proposals made fairly easy targets for activists. Winter navigation and out-of-basin diversions were issues that created a sense of shared regional interest. Both involved perceived future threats with potential costs throughout the entire Great Lakes system that still could be averted by proactive cooperation. Both had larger-than-life 'bad guys' – sunbelt speculators and the Corps of Engineers – neither of which had the ability or intention to act immediately on their proposal. Nothing was inevitable about the proposals. The various engineering schemes promised future, highly speculative profits. Despite the proposal's simple surface logic, the cost-benefit considerations were ludicrously out of balance. A variety of woes had befallen the Great lakes shipping industry, making it inconceivable that the economic benefits in increased shipping and toll receipts could ever approach the engineering maintenance costs required for winter navigation. The proposed schemes reeked of pork barrel politics and had little overt political support, even from the leaders of portside communities. They represented, therefore, no entrenched powerful economic forces at work, no workers to be displaced, and were, as a result, good organizing targets.

A second perceived threat involved proposals to divert Great Lakes water beyond the basin boundaries to dry regions of the United States. A variety of engineering schemes have been proposed at one time or another to use Great Lakes water to irrigate midwestern agriculture, to move western coal via a coal-slurry pipeline, and, most recently, to raise the Mississippi made shallow by drought. The threat of large-scale diversions was among the first concerns of Canadian and US negotiations that led to the Boundary Waters Treaty in 1909 and has had the effect of highlighting the mutual economic interests of the Great Lakes region. Great Lakes officials began to see their abundant supplies of water as a competitive advantage against the so-called sunbelt (which some editorial columnists in the Great Lakes region had come to call the 'parchbelt'), and talk of tapping into the Great Lakes supply were fighting words. The economies of the western United States and Canada have been stimulated by oil and mineral production, irrigation agriculture, and tourism; at the same time the Great Lakes region has suffered a steady decline in its heavy manufacturing-based economy. If the availability of fresh water was to become a limiting factor for recently expanding economies of the sunbelt, then the Great Lakes, the world's largest supply of surface liquid fresh water, could one day be the source of more wealth than all the oil in Texas. With such visions in mind, the states bordering the lakes eventually formed a Great Lakes Charter, agreeing to consult with each other before any significant diversions would be allowed.

Wayne Schmidt, a staff ecologist with Michigan United Conservation Clubs (MUCC), recognized the difference between natural coalition-building issues like winter navigation and other more difficult questions:

> Winter navigation was a natural issue which brought all the entities together. But things aren't always so clear cut. It's difficult to get people in Quebec and Wisconsin to get together on water quality issues. This federation [that became GLU] is a gamble, but we're going to give it a try.[55]

From the beginning there were disputes over the most important issues from an environmental perspective and those issues most suitable for building broad coalitions. Many of the environmental groups involved in the Great Lakes at the time, including Lake Michigan Federation, Sierra Club, Operation Clean Niagara, and others, had the toxics issue clearly on their agenda. From the perspective of organizing regional cooperation among environmental advocacy groups, however, winter navigation and diversion had several advantages over the more complex issues of toxic contamination. The issues of diversion and winter navigation, unlike issues of toxics, were variations on century-old debates regarding management of public resources.[56] Positions of the actors could be defined and variously interpreted according to the terms of those experiences. Toxics, on the other hand, had involved physical and political factors that complicated advocacy groups' organizing strategies. These include the fact that:

1 Because toxic substances get into the ecosystem as the by-products and waste of essentially every current major economic activity, the 'bad guys' are not distant schemers or government bullies, but are all around us. They are difficult to locate but, when identified, make powerful opponents.

2 Because toxic contamination is so ubiquitous and its sources so diffuse, measuring its effects is complicated by the absence of an uncontaminated 'control' population or a 'quiet' background against which to measure the toxic 'noise'. It is extremely difficult to locate specific effects on human populations that can be directly attributable to specific toxicants.

3 Quantities of toxics, although immense in the aggregate, are highly diffuse

and diluted in immense volumes of water. The costs of cleaning up any single source is usually far greater than the benefits to be derived from any individual clean-up when that clean-up's benefit is measured as a proportion of the overall problem.

4 Because clean-up is so costly and inefficient, it is better to prevent pollution in the first place. But pollution prevention requires process changes and lifestyle adaptations that go to the heart of personal choices and economic processes. As a result, pollution prevention may not be the best ground on which to organize advocates across class, race, and cultural lines.

5 Finally, organizing around toxics issues is further complicated by the threat posed to the hunting, fishing, and tourism industries by widespread public fear of toxic contamination. These industries form the economic underpinning of the hunting and fishing clubs such as MUCC. In addition, the influence of the United Auto Workers in MUCC at the time and later in Great Lakes United cannot be discounted.

Thus, one obvious organizing strategy – increasing political pressure by tapping public fears and concerns about contamination – might have proved very costly to a major component of the proposed coalition. Hence, although government's difficulties in addressing the issues of toxic contamination opened the door for environmental advocacy, the complexities of the issue also threatened to undermine the capacity of environmental NGOs to attract broad constituencies.

These issues have bedevilled those who have tried to organize around toxic pollution. Despite these complications, however, improved understanding of toxics led to growing acceptance by the scientific community, governments, and advocacy groups of the necessity of an ecosystem approach to stopping pollution. A major strategic shift resulted from the realization that many of the localized problems had distant sources and a common thread – the degradation of the Great Lakes ecosystem as a whole. This shift was a recognition that the individual components of the Great Lakes ecosystem function together and that actions in one part may have unpredictable ramifications in another. This recognition occurred in different ways for different organizations and regions, but developed parallel to the articulation by the IJC and the Canadian and US water quality agencies of the need to take an ecosystem approach to handling Great Lakes pollution. To assert political power at the ecosystem level, the focus of environmental advocacy needed to be at the level where decisions were made that affected the entire ecosystem.[57] There was a growing sense among scientists and policy experts, the 'epistemic community' of the Great Lakes, that the existing institutional structures were incapable of resolving the crises facing the ecosystem. The academic and scientific meetings of these scientists took on an increasingly political tone. It was in this context that many of Great Lakes environmental interest groups saw the need to form an alliance despite their differences. The effort to create such an alliance constitutes the early history of Great Lakes

United and demonstrates many of the forces that hold environmental coalitions together, as well as some of those that tend to rend them asunder.

Coalition building

Tom Washington, the dynamic executive director of Michigan United Conservation Clubs (MUCC), a statewide coalition of sports and conservation
clubs with nearly 200,000 members, was among those who believed that
some kind of regional Great Lakes federation was needed. With staff
ecologist Wayne Schmidt, he conceived in 1981 a Great Lakes organization
based on the MUCC and National Wildlife Federation models of a coalition
of like-minded organizations with a strong central administration to address
shared concerns across the basin.

In MUCC's first press release on the matter, issued in November 1981,
Washington said he planned to establish a federation to protect and improve
Great Lakes water quality:

> This federation could be instrumental in the long-term protection and
> improvement of Great Lakes water quality through citizen action. It could
> be a valuable tool in educating citizens and organizations in the Great
> Lakes basin about the inter-relationship of the waters of the basin and the
> need for an 'ecosystem approach' to managing water and other natural
> resources of the Great Lakes.[58]

With a grant from the Joyce Foundation, the MUCC staff began the process
of bringing Great Lakes organizations together. The debates and manoeuvring that ensued reflected in many ways the nature of the problems being
addressed. The perception of common threats and mutual interests brought
people together; fears that their individual interests, styles, and philosophies
would be subsumed by a dominant central authority drove them apart. Mixed
with political disagreements and differences in organizational styles was the
involvement of several controversial and flamboyant individuals.

Organizational structure and leadership

Differences in political styles and goals nearly scuttled early efforts to form
a Great Lakes coalition organization. In the summer of 1982 MUCC and the
Joyce Foundation issued invitations to the leaders of Great Lakes organizations to meet at Mackinac Island near the straights dividing Lakes Huron
and Michigan. Fifty-five delegates, from eight states and two provinces,
attended. The debate quickly centred on what kind of organization should be
created. Probably the most critical dilemma facing the new coalition was the
contradiction between the decentralized nature of a diverse coalition and the
need for strong leadership to hold the diversity together and represent their
many interests. Tom Washington and others argued for the formation of a
strong regional organization that could advocate positions with a single voice

representing the scores of groups with environmental portfolios. Many others saw a need for a central information clearing-house and networking node for existing groups, but feared a new organization would compete with them for influence, funding, and members. The issues of organizational structure were mirrored in leadership styles. Tom Washington and Wayne Schmidt worked in and were used to authoritarian, decision-making styles, whereas many of the environmental organizations involved early in the Great Lakes coalition-building promoted a more egalitarian, participatory style.

The Mackinac meeting ended inconclusively, with the issuing of a consensus document focused on the need for vigilance against the threat of diversions and concerns about pollution and the agreement to form an organization whose structure and purpose were to be hammered out at a second meeting six months later in Windsor.

Organizational and personal issues came to a head in Windsor in November. One hundred and ten delegates representing seventy groups from all the Great Lakes states and provinces agreed to form a coalition, but it ousted from leadership roles MUCC's Tom Washington and Barry Freed/Abby Hoffman, the most controversial figure in GLU's early history. The decisions made regarding leadership and organizational structure, as well as the skills honed in securing them, helped establish GLU credibility and legitimacy in the basin.

Leadership

Abby Hoffman had been a leading organizer of many of the highly publicized displays of 1960s radicalism and a key figure in the loose network of theatrical protesters known as the yippies. He had an uncanny ability to broadcast his brilliantly succinct political symbols by capturing the attention of America's news-entertainment complex. Out of his ability to manipulate the national media he crafted a unique political philosophy and strategy, which he preached enthusiastically. Convicted but later exonerated for incitement to riot in his role as one of the leaders of anti-war demonstrations at the 1968 Democratic National Convention in Chicago, he went underground in 1972 to avoid cocaine trafficking charges which he insisted were part of a frame-up. During his years in hiding he created, with plastic surgery and short hair, the persona of Barry Freed, a St Lawrence resident and leader of the Save the River environmental organization. In 1980 he surrendered to US authorities and was released from prison in spring 1982. As a representative of Save the River, he joined the meetings at Mackinac and Windsor.

Hoffman's active role in GLU's founding and the media attention he attracted were more than many of the established sporting groups could stand. Ironically, Hoffman was linked in most of the news accounts with MUCC's Tom Washington as the pair that argued most forcefully for an activist organization. MUCC, a staid conservation and sporting organization, was scarcely a club for former yippies, and its leadership was more than a little uncomfortable with its new ally. Thus, MUCC's role as leader of the new coalition became increasingly complicated.

Representatives of the National Wildlife Federation, Sierra Club, Audubon, Toronto's Pollution Probe, the binational education group Great Lakes Tomorrow, League of Women Voters, the Federation of Ontario Naturalists, Lake Michigan Federation, Sierra Club, and others joined in Windsor to reject a strong executive director and an independent board of directors, favouring instead a decentralized organizational model structured around 'task forces'. The idea of task forces was that each would focus on an issue and be made up of representatives from member organizations interested in that specific issue. The task forces would recommend political strategies to the coalition and sometimes carry out activities in the name of the coalition. The task force structure would prevent a centralized authority from making decisions on behalf of the many local members. The *Detroit Free Press* quoted Mimi Becker, Great Lakes Tomorrow project manager, as saying, 'We won't have action done by some executive director that goes running around from state to state. The political action and the credibility must be implemented by local groups.'[59]

Organizational model

There was concern that organizations with large memberships, such as the 100,000-member MUCC, would crowd out smaller groups; that those organizations with hierarchical structures, such as the labour unions, could dictate solid block votes; and that groups such as Save the River, with media stars like Hoffman, could end up speaking for the group. There were also strong concerns about gender equity and male-dominated leadership. Many of the groups opposed to the centralized structure were represented by competent, politically astute women leaders: Glenda Daniel at Lake Michigan Federation, Carol Swinehart with the League of Women Voters, Jane Elder of the Sierra Club, Pamela Chase of Pollution Probe, and Mimi Becker of Great Lakes Tomorrow. They were suspicious of the organizational leadership styles brought to GLU's founding by Washington and Hoffman. Furthermore, concerned Canadians feared that US groups would dominate. The US groups had initiated the coalition-building process, obtained the initial grant money, and were already larger and more powerful than their Canadian counterparts. These concerns were played out in the structuring of the by-laws. The debate over whether individuals or organizations should have membership and voting privileges was really about whether large organizations, like the 100,000-member MUCC, could dominate. The debate over the role of an executive director was really a referendum on the personalities and styles of Washington and Hoffman.

Those favouring a strong executive argued that environmental problems often required quick responses and that some central authority needed to be empowered to act without the lengthy procedure of setting up a task force for each problem. Proponents of a decentralized structure, on the other hand, pointed to the Clean Air Coalition and the coalition of groups that had fought the Alaskan oil pipeline as examples of coalitions that still managed to respond quickly and were more successful than authoritarian groups in arousing grass-roots support. The decentralists won the day, and by the end of the Windsor meeting the founding members of the new organization, Tom Washington and Wayne Schmidt and their activist ally, Abby Hoffman, had been rebuffed. Washington and Hoffman threatened to pull their organizations out of the coalition. Washington was quoted in the *Bay City Times*, 'I don't think we'll be really active participants in this organization. The amendments [passed at Windsor to reduce central authority] take away any strength or power to act in an expedient manner.'[60] Abby Hoffman, quoted in the *Detroit Free Press*, was, as usual, more to the point: 'We are interested in political action. We're sick of groups that sell newsletters.'[61]

In the six months between the November 1982 Windsor meeting and the first annual meeting of Great Lakes United in Detroit in May 1983, the mainstream conservation groups were able to agree to a set of by-laws and policy statements. They also agreed on a president, Bob Boice, who was a career employee of the New York State Department of Environmental Conservation and an officer in the New York State Conservation Council, a 350,000-member coalition of sporting clubs and National Wildlife Federation and National Rifle Association affiliate in New York. In personal style, they could not have found someone more removed from Abby Hoffman's style than Boice, who was regarded as a consensus-builder.

The first GLU meeting adopted a series of policy resolutions. In addition to taking stands against toxic pollution, winter navigation and diversions, the group declared its support for a US–Canada Air Quality Agreement modelled on the Great Lakes Water Quality Agreement and meant to eliminate acid rain; for acquisition and improvement of national park lands in the Great Lakes, and opposition to the Reagan administration's Interior Department park policies; for renewal of US clean water and clean air acts; and for increased funding for Great Lakes research and water quality monitoring.

Strategizing and gaining leverage

Between the first annual meeting and the review and renegotiation of the Water Quality Agreement in 1987, Great Lakes United grew steadily in numbers and influence. By 1986 membership had grown to more than 200 diverse groups and hundreds of individual members from the United States and Canada 'striving for proper management and protection of the Great Lakes and St Lawrence River'. A full-time executive director and support staff were hired, and headquarters were established in Buffalo, New York, at Medaille College. Plans were underway to open a Canadian office in Windsor, Ontario.

Strategy

As noted earlier, the Great Lakes Water Quality Agreement was scheduled for a formal review in 1987. The leadership of Great Lakes United saw this as an opportunity to draw attention to the agreement's principles, which they believed were still being largely ignored by the parties. Because GLU was a relatively new voice in the basin, its officers had been considering a 'tour' of the lakes, in the form of public meetings throughout the basin similar in scope and intent to the PLUARG meetings.[62] The idea for a tour originated in 1985 as a way to promote membership in GLU. By the following year the concept converged with the officers' concerns about the upcoming scheduled review of the agreement. There seemed to be reason to worry about the fate of the agreement under review: the Reagan White House and Mulroney's Progressive Conservative government in Ottawa were perceived to be hostile to federal action on behalf of the environment – the kind promised by the agreement.

John Jackson, a veteran organizer around issues of toxic waste dumps in the province of Ontario and a Great Lakes United activist, suggested that GLU, rather than just doing a promotional tour, should organize a series of 'public hearings' around the basin to gather testimony regarding the government's progress in implementing the agreement. The hearings were intended to raise GLU's profile in the region and to build a base of support for the principles and goals of the agreement,[63] pre-empting and prompting US and Canadian government officials who had still not discussed publicly plans for the upcoming agreement review.

The GLU board of directors approved the concept of the citizen hearings and established a Water Quality Task Force under the GLU by-laws. The task force sought and received funding from the C. S. Mott and Joyce Foundations. GLU also hired a Water Quality Task Force coordinator, Tim Eder, who later became one of GLU's representatives on the US team for the agreement renegotiations. Before joining GLU, Eder had worked with Save the River in upstate New York.

Eder and GLU organized nineteen 'Citizen's Hearings on Great Lakes Water Pollution' in cities across the Great Lakes basin. GLU estimates that more than 1,200 people attended, of whom 382 made statements at the hearings or sent in comments by mail. Members of the GLU task force received testimony from residents and joined with local organizations and reporters on tours, often in boats, to witness environmental problems in their area. They saw open piles of coal, salt, and scrap iron lining the banks of the Milwaukee harbour, all together feeding lead, chromium, mercury, arsenic, and phenols to the harbour's already contaminated sediments with every run-off event. In Green Bay they toured paper mills where an estimated fifty to seventy pounds of PCBs per year were legally discharged in the pulp mills' waste water. From their boat, GLU task force members were shown several waste dumps on the shores of Lake Superior at Duluth. In Massena, New

York, the group toured the St Lawrence River near the General Motors and Alcoa industrial waste sites, where they heard a New York Department of Environmental Conservation official describe the 'contaminant plume of considerable proportion migrating from the industrial landfill and discharging to the St Lawrence River'.[64] These scenes were repeated throughout the nineteen stops on the tour. The testimony gathered was emotional and dramatic: 'What we pump down the sewers this week will end up in our cornflakes next year and eventually in my blood and fatty tissues', argued one witness in Toronto. A Native American leader in Cornwall, Ontario, said that Native American families used to eat twenty to thirty pounds of fish every week, but now they warn children and women not to eat any fish because the flesh is contaminated. A deformed cormorant found locally sat at the hearing table in Green Bay. In Montreal, biologists studying the decline of Beluga whale populations in the St Lawrence estuary reported finding whale carcasses coming ashore with high concentrations of PCBs, mirex, and dioxin in their flesh.[65]

Taking a stand: GLU's position

From the citizen hearings, GLU compiled *Unfulfilled Promises: A Citizen's Review of the International Great Lakes Water Quality Agreement*. The Water Quality Task Force recommended that:

1 Governments seriously act on commitments to end the release of persistent toxic substances into the lakes,
2 Governments and the IJC better inform the public on water quality issues and involve citizens in all levels of water quality decision making,
3 More research be done on human health effects of toxins at levels found in the lakes,
4 New chemicals be tested for toxicity and persistence before they are allowed to be manufactured and used,
5 Methods be improved for eliminating pollutants in-place in contaminated sediments,
6 The practices of overflow dredging and open lake disposal of contaminated dredge spoils be ended,
7 The IJC become more active in commenting on government programmes and involve the public in all aspects of the IJC work.

The very first recommendation in the report commented on the scheduled GLWQA review. According to the report:

Most of those who spoke at the hearings emphasized the need for immediate actions to correct water quality problems. Renegotiation of the Great Lakes Water Quality Agreement at this time would mean the diversion of resources and a resultant delay in addressing these problems. In addition, many speakers expressed little faith in the Reagan and

Mulroney administrations' commitment to protecting the environment. They feared that if opened up for renegotiation at this time, the Agreement would be weakened.

These two concerns combined with statements from all Parties at the hearings that the Agreement is a document that encourages positive actions and does not discourage such action, leads the GLU Task Force to conclude that the Agreement should not be renegotiated now.

Therefore, the GLU Task Force recommends that the two federal governments not renegotiate the Great Lakes Water Quality Agreement at this time.[66]

GLU's resistance to tampering with the agreement echoed the sentiments of the participants at the hearings, as well as the conclusion drawn by a study committee of the Royal Society of Canada and the National Research Council of the United States (RSCNRC) which had reviewed the agreement in 1984.[67] Both the scientific and environmental advocates reviews concluded that the GLWQA was fundamentally sound and required determined implementation by the parties, rather than renegotiation and amendment.[68] Both reports were made widely available to governments and the media. According to Ron Shimizu, who at the time was responsible for Environment Canada's implementation of the GLWQA, GLU's emphasis on the positive features of the agreement, which must not be tampered with, 'set the tone, the public parameters of acceptability around which the governments could conduct a review'.[69]

Party positions

Despite the GLU's anxiety that opening the agreement to negotiations at this time could be a prelude to disaster and that the agreement could be greatly weakened, the individual environmental officials in charge of the review – Ron Shimizu, Canada, and Kent Fuller, United States – were committed to the basic framework of the agreement and had communicated that sentiment to each other.[70] Both believed that the GLWQA could benefit by two kinds of amendments: technical changes that would specify government commitments to respond to emerging pollution problems and management changes that would clarify roles and make the governments more accountable for their commitments by linking the goals of the GLWQA more closely with specific programme elements.

Although most of the public comment and review centred on specific pollution-related problems, the bureaucrats charged with preparing their governments' positions for the negotiations were also concerned about water quality management issues, specifically the vague distribution of responsibilities between the IJC and the governments. EPA's Kent Fuller wanted to modify the GLWQA so that its lofty goals would be explicitly related to federal water quality programmes and commitments in each country. As the agreement stood on the eve of its review, goals were established and the IJC reported on progress, but there were few practical mechanisms to tie the goals to particular programmes. This lack of accountability further complicated an already daunting evaluation task.[71]

The binational programme with the highest profile, and the one of most concern to many of the people who had spoken at the Great Lakes United hearings, was the Remedial Action Plan (RAP) programme initiated by the IJC Water Quality Board in 1985.[72] The programme called for clean-up plans to be developed for each of forty-two locations in the United States and Canada, mostly heavily polluted rivers and harbours, designated 'Areas of Concern'. The RAP programme had begun to take on a life of its own, adopted by local community activists and regional environmentalists, as well as some state environmental agency personnel who saw it as giving new life through the stimulus of international attention to the effort to clean up some particularly entrenched pollution problems. The RAP programme, because of participation by state agencies and scores of community activists as well as the IJC, was approaching an *ad hoc* institutional status by 1987, although it did not have standing in the Water Quality Agreement, the Boundary Waters Treaty, or domestic law. By incorporating the RAPs into the agreement, the parties' responsibilities for preparing them would be made clearer.

Remedial Action Plans

It is necessary to consider at this point the history and recent evolution of the RAP programme because in many ways the process of developing and implementing RAPs reflects trends in the relationships between non-governmental organizations, intergovernmental organizations like the IJC, national governments, and the state, provincial, and local jurisdictions.

The Remedial Action Plan (RAP) programme marks a departure from the traditional IJC activities. In the past the IJC made recommendations only when both governments asked it for specific studies. The type of recommendations and the range of information expected by the governments was clearly delineated in the reference, or request, issued jointly by the two national governments. Since both nations had to agree before any issue could be referred to the Commission, referenced issues tended to be those in which mutual interests were considerable and obvious. Care was taken not to suggest any threat to national sovereignty. In fact some observers have credited the commission's steadfast avoidance of bilateral conflict through the reference procedure for its longevity and success.[73] Before the RAP programme, the IJC acted for the most part only at the behest of the federal governments. An axiom commonly used to paraphrase this relationship was 'the governments do, the commissioners review.'

The RAP programme was a departure in that the IJC, through its Water Quality board, was, in effect *directing* the parties to develop clean-up plans.

The IJC determined what should be in a RAP, what criteria by which the RAP would be evaluated, and what constituted adequate public consultation and citizen participation. In addition, individual IJC staff members, by becoming involved in local watershed planning, were developing professional, personal, and political relationships with community activists and with the local and state officials who were charged with producing the RAPs. By becoming involved in water quality planning processes at subnational levels, the IJC risked being perceived by the parties as overreaching its mandate and meddling in sovereign affairs.

The IJC found itself in a conundrum. The severe, but localized, contamination in the areas of concern posed a limit on further progress in Great Lakes clean-up. Unless water quality in these areas was improved, the objectives of the GLWQA were unlikely to be achieved. But the areas of concern were by definition local problems, requiring local efforts and investment to remediate. How could the IJC, an international body responsible to the federal governments, invigorate local communities to take responsibility for their piece of the ecosystem? It seemed crucial to involve as many influential parts of the community as possible. In its third biennial report issued in 1987, the IJC recommended that 'all levels of government take steps to foster community support and involvement in developing and implementing the remedial action plans.' According to a pamphlet on RAPs published by the IJC,

> Each citizen can play a valuable role in the RAP process, by contributing information on Areas of Concern and providing support for the development and implementation of the plan. This endeavor can only be successful if a concerned public is involved in developing and implementing each remedial action plan.[74]

By encouraging public participation in remedial action plans, the IJC was, in the environmental politics of the late 1980s, necessarily becoming involved with environmental activists and activism. It was carving out new relationships between jurisdictions and citizens in a previously unprecedented way.

The IJC was not necessarily interested in participatory democracy as an end in itself, nor as any principled political position,[75] but as a means to gain support for its preferred policies, those based on an ecosystem approach to RAPs. According to the IJC, 'each RAP must embody a comprehensive ecosystem approach to restoring and protecting beneficial uses in the area of concern'. The kind of public participation sought by the IJC officials promoting RAPs was the sort provided by an involved constituency. The burden of promoting the RAPs, therefore, fell to local environmental activists with recognized interests and credibility and who also shared a personal sense of responsibility broad enough to include the Great Lakes ecosystem as a whole. These describe, in fact, many of the local activists involved in Great Lakes United member organizations and other environmental NGOs. To make the RAPs work as intended, therefore, the IJC finds itself in alliance with local environmental activists and doing so at the risk of alienating some of the very state and local officials responsible for delivering a RAP to the IJC as well as other local economic and business interests.[76]

The commission's difficulties resulted from a contradiction inherent in the RAP programme and its ecosystem approach: that the benefits accrue to the entire ecosystem, while the costs are borne disproportionately at the local level. The RAP programme was intended to make progress in polluted areas of concern that consistently failed to meet the water quality objectives of the agreement. To achieve the ecosystem-wide goals, therefore, hot spots have to be addressed. The ecological significance, then, of some of the hot spots is primarily in their basin-wide impact. The political significance, however, is in the fact that the local community is being asked to shoulder the responsibilities of planning and implementing a clean-up programme that is likely to be very expensive and which may fail to result in any substantial local benefit. Furthermore, many of the hot spots are heavily industrialized or otherwise degraded areas that lack a politically powerful resident constituency. The IJC-designated areas may compete as well with other environmental problems in the local area, some of which might be more visible or have more direct impact on the local community. The success of many of the RAPs may depend in the end on the IJC's ability to build environmentally astute community advocates in the RAP areas. Such a condition for success for one of the IJC's most cherished programmes could easily lead to government efforts to rein in the scope of the commission's activities.

This tug and pull between the governments and the IJC was nothing particularly new. Ever since its first reference reports on water pollution early in the century, commissioners have occasionally recommended increased authority for the IJC. In 1981 the commission proposed a new, expanded role for itself beyond its traditional role as scientific and technical advisor to the governments,[77] suggesting, in fact, something like the role it adopted for itself in the RAP programme. This recommendation was rejected by the Office of Canadian Affairs in the US State Department, which told the IJC that 'rather than a broadening of the Commission's Great Lakes focus as proposed, the State Department believes that the Commission should continue to devote its efforts with greater precision to the technical questions specified in the 1978 Agreement.'[78]

The NRCRSC report was also critical of the IJC's expanded role, in particular the Water Quality Board. The report recommended that 'the coordinating responsibilities for the control programs that implement the Agreement be left to the Parties, rather than to the Water Quality Board. This coordination should be handled through bilateral government-to-government meetings.'[79]

According to Munton, the purpose of the NRCRSC criticisms was not to reduce the commission's importance, but to increase its independence and, therefore, its effectiveness. Governments should be clearly responsible and accountable for the commitments they make under the agreement.

By the time of the 1987 review of the GLWQA, the RAP programme had become one of the most active water quality efforts in both countries. Yet before the 1987 Protocol, the programme itself was not institutionalized in the body of the GLWQA. This lack of institutionalization meant that a major Great Lakes anti-pollution effort was outside the official agreement framework, a kind of rogue influence, indeed posing serious structural and political challenges to the water quality bureaucracies in both countries and all the states and provinces. The RAP programme bore the burden of relying almost exclusively on popular political support for its legitimacy. Professionals associated with the programme wanted to ensure government accountability for the RAPs and to tie RAP programmes to specific and measurable end-points to strengthen its legitimacy and raise its ranking on government priority lists.[80] The RAP situation added to the sense that management functions under the GLWQA needed clarification, and that the specific responsibilities for developing and implementing RAPs be expressed and embraced by the governments through the agreement. It was among the goals of both the Canadian and US governments, therefore, to add specific language to the Agreement to formally incorporate the RAPs into the GLWQA. With such language the parties would be able to reassert their authority over the planning process, even if the annex accommodated fully the original IJC guidelines.

Great Lakes United concerns

By reviewing the GLWQA, GLU advocates expressed complex and contradictory opinions on the role of the IJC. On the one hand, the IJC provided a focal point for the many water quality activities in the basin. Its mandate from the 1909 Boundary Water Treaty preceded and was not unlike what GLU understood its own to be: looking out for the interests of the entire watershed against the parochial regional and national interests throughout the basin. On the other hand, the IJC had no authority to take meaningful regulatory or clean-up action, and few financial resources. Responsibilities that were left to the IJC, it was feared, would be those the governments preferred to avoid. GLU and other NGOs preferred an IJC that could and would prod the governments, applying, where appropriate, the pressure of public concern to goad the parties, the states, and the provinces into action.[81] These contradictory opinions were never really debated or resolved in the strategy deliberations within the organization. Instead, most of the effort was put into achieving the goal of protecting the agreement from what they feared might be cynical manoeuvres from the conservative governments in power in both administrations. For the most part the IJC was perceived by GLU strategists as another governmental institution to be lobbied. The goal of GLU's lobbying was to convince the commissioners to take an activist stance in promoting environmentalist positions before the governments. The fact that as a treaty organization, the IJC was a creature of governments and unlikely to be successful as an independent activist organization was of little concern to the activists.

The stance taken by both parties in preparation for the 1987 review was not publicly critical of the IJC. Their position instead was that accountability and management for agreement activities and responsibilities needed to be made explicit. The review and renegotiation provided an opportunity to clarify roles and responsibilities. Still, although representatives of several government agencies and the states and provinces (including Quebec, which is not part of the agreement or the Boundary Waters Treaty) participated with the citizen group representatives and the US State Department and Canadian Ministry of External Affairs, the absence from the negotiations of any representatives of the commission, even as technical advisors, is note-worthy.[82] And the result of the 1987 agreement as eventually adopted, as discussed later, reasserted the primacy of the parties over the IJC in agree-ment activities and placed into question the future of the IJC as an effective binational organization.[83]

THE 1987 REVIEW AND RENEGOTIATION OF THE GLWQA

GLU's citizen hearings and the widely publicized review of the GLWQA released by the NRCRSC raised the agreement's profile and made clear to the US and Canadian environmental agencies the existence of a vocal and organized constituency that supported the agreement's purposes. As a result, beginning early in 1987 both sides went to unprecedented lengths to include the public from the beginning stages of the process. Shortly after the parties began preparing draft position statements and proposed amendments, they consulted with a range of public representatives. In Canada, both the federal and provincial environment agencies, including their top administrators, held meetings with citizens groups to discuss the agreement. Even preliminary drafts of amendments the Canadian government was considering proposing to the American side were given to key Canadian environmentalists for review and comment. Environment Canada and the Ontario Ministry of the Environment co-hosted an open workshop in July that was attended by scientists and labour representatives, government agencies, industry associa-tions, lawyers, environmental groups, native groups, and educational institu-tions. Advanced materials were circulated with discussions of possible amendment areas, including RAPs, groundwater contamination, airborne pollutants, contaminated sediments, and research needs. After the workshops, Environment Canada and Ontario Ministry of the Environment held public meetings in Kingston, Windsor, and Sault Ste Marie.

The United States also provided opportunities for public comment on the American positions but were, in general, less well organized, and fewer people attended the public meetings than in Canada.[84] Draft positions were widely circulated, however, both in an original amendment form and, responding to the widely expressed opinion that changes should only be made to the annexes, in a second, all-annex version. The draft amendments were also circulated for comment to various interested federal and state agencies.

A US caucus was organized to derive a position for negotiations with the Canadians. It included representatives of the EPA, Coast Guard, Office of Management and Budget, Army Corps of Engineers, National Oceanic and Atmospheric Administration, Department of Agriculture, and Fish and Wildlife Service. Each of the eight Great Lakes states was involved through representatives appointed by the governors. Three of them were chosen by the others to represent all the states as participants in the US caucus and in the binational negotiations. According to a summary of the review process prepared by EPA, review of the agreement quickly reached consensus on five conclusions which served as assumptions throughout the process of drafting amendments.

- Existing Agreement is basically sound.
- The purpose and general goals and objectives must not be changed.
- It would be desirable to bring the Agreement up-to-date.
- It would be desirable to tighten accountability and management.
- Review and amendment must be completed quickly to avoid diverting resources from implementing the existing Agreement.[85]

After it became clear to the leadership at Great Lakes United that the governments intended to proceed with some changes to the agreement, GLU's leaders shifted strategy. They no longer insisted that the agreement remain untouched, although they continued to express that preference. But they also urged that, if the agreement were to be altered, formal amendments to the body of the agreement be avoided by placing all changes in annexes and, if changes were proposed, they should be considered with the full participation of the public. A memo dated September 10, 1987, from John Jackson and Tim Eder, GLU leaders, addressed to 'People Interested in the Future of the Great Lakes Water Quality Agreement' summarized the position GLU had presented to US and Canadian officials and explained that

> The reason we insist on new Annexes only is simple: to protect the strong provisions of the existing Agreement, such as zero discharge and virtual elimination of persistent toxic substances, from being weakened. We believe that the only thing that should be on the negotiating table is new Annexes or supplements to existing Annexes.

> . . .The public should have a major role in the actual negotiations. GLU has petitioned officials in both countries to grant representatives of organizations such as GLU 'observer status' in the negotiations.

Central to GLU's political principles has been a belief in environmental advocacy through participatory democracy. This belief rests on the assumption that unless meaningful public participation is broadly encouraged, the only interests represented and articulated in the decision-making discourse will be those with the most at stake financially and professionally: regulated industries, the polluters (especially past polluters, the so-called responsible parties), their consultants, and the professional staff of the environmental agencies. According to this view, even when government agency personnel are inclined to defend the public's health and welfare and their right to a healthy environment, they are often overwhelmed by the short-term logic of economic expediency. The instinct of the civil servant to represent broader public interests has to be supported and encouraged and, where absent, demanded. This, GLU believes, is one of the roles of the environmental advocate in the decision-making process.

Even when governments undertake public participation and public consultation activities, they often do so in a manner that suggests government's responsibility is to strike a balance between competing stakeholder interests, as though all stakeholder interests were of equal value and each had equal power, ability, and motivation to articulate and defend its interests. Yet stakeholder rights and interests are multidimensional and power is not equally distributed, nor are costs and benefits. Furthermore, in the view of environmentalists, the right to pollution-free waters should be given inherently more weight than the right to use those waters to discharge wastes. One way for a government to clarify rights, responsibilities, and overriding interests is to articulate in a public document a set of principles and goals to which that government is committed. This kind of statement says these are overriding principles and, where they threaten certain special interests and privileges, the principles ought not to be weakened to accommodate them. The Great Lakes Water Quality Agreement is just such a document, and as an international agreement, it has the additional authority created by the history of bilateral relations and mutual treaty obligations.

GLU recognized the importance of the GLWQA as a statement of principles and placed significant organizational emphasis on defending it. When GLU first approached administrators in EPA and Environment Canada with its request to be part of the binational review and amendment process, it was a bold, if understandable, move. Direct participation by NGOs in binational affairs was rare, but not without precedent. Mark Van Putten, a lawyer and director of the National Wildlife Federation's Great Lakes Natural Resource Center knew of a migratory waterfowl treaty between Alaska and Canada in which certain sporting groups affiliated with the Wildlife Federation had officially participated.[86] According to Tim Eder, it was a crucial realization, and, based on Van Putten's example, GLU began to insist with increasing confidence that the governments give GLU's representatives seats at the negotiating table.[87]

GLU's case was straightforward. Its leaders had always demanded that citizens be part of the decision-making process. Now that decisions were being made at the bilateral level, their demand would be the same. They sincerely believed that as a result of the success of their citizen hearings they had earned the right to represent and advocate broad citizen interest in the negotiation. They felt obliged to the public who had testified at citizen hearings to make sure public opinions regarding the agreement were heard.

They believed that they and other environmental advocacy groups represented interests shared across the international boundary, interests that were broader or more fundamental than specific national interests, the interests of what has been referred to as the Great Lakes constituency. This formulation is, of course, inherent in GLU's existence as a binational citizens group whose identity (even, perhaps, its definition of homeland) consists of an ecological region with watershed frontiers rather than political borders. GLU's participation in the bilateral review and negotiations would, therefore, represent a fundamental (even if minor at this point) challenge to the legitimacy of the nation-state system for issues of environmental protection, even if the parties involved did not necessarily see it in that way.

Once GLU had decided to petition officials of both governments for representation, according to GLU's Tim Eder, they, 'had to pull out all the stops and play every card in our bag to get a seat at the table, that didn't just happen.'[88] The campaign to gain observer status began in June 1987. GLU wrote letters to the foreign ministries and environmental agencies of both countries. For several months they simply received no answer, not even an acknowledgement of their request. GLU let both governments know that they were not going to let up until an answer was received.

One decisive tactic was that all the NGOs represented (Sierra Club, NWF, and GLU) had built alliances with many Great Lakes representatives in Congress and the Senate. GLU contacted congressional supporters, particularly Congressmen Overstar (D–Michigan) and Nowak (D–New York) who had been leaders in Congress in several Great Lakes issues, asking them to intercede with the State Department on its behalf. In addition, the Senate Great Lakes delegation sent a letter to Secretary of State George Schultz, signed in June by all the senators in the region, which referred to GLU's citizen hearings in which 'a lot of good ideas' were gathered and suggested to the State Department that it should consider GLU's appeal. Another letter was circulated among the Northeast-Midwest congressional delegation. That letter stated that 'we recommend that some community-based citizen group be given observer status at these discussions.'[89]

GLU also leveraged the positions of the two federal environmental agencies. Ron Shimizu of Environment Canada told Eder and Jackson that his agency did not have a problem with GLU's request for observer status, but he believed the US side would not agree. Then, in discussions with EPA's Kent Fuller, Eder and Jackson reported what Environment Canada had said, hoping to encourage Fuller to support their request.[90] When GLU appealed directly to the US State Department, according to Eder, officials there never discussed environmental issues but instead expressed foreign policy concerns and their desire not to complicate US–Canadian relations by involving the NGOs in the negotiation. A parallel lobbying effort was undertaken in Canada. Canadian GLU representatives met with Canadian Environmental Minister Tom McMillan as well as Elizabeth Dowdswell and Ron Shimizu, senior bureaucrats in Environment Canada.

In the end the campaign was successful. According to EPA's Fuller, who was on a full-time assignment to develop EPA's position for the GLWQA review, the NGOs had successfully convinced him and others within the EPA that they were interested in 'cooperating, not disrupting'.[91] The personality of Fred Jones Hall, the State Department official given responsibility for the GLWQA review, also played a major role in sanctioning NGO participation. Hall was new to the department, a successful Texas businessman with a can-do attitude. He advised EPA members of the US delegation that he would rely on EPA's technical advice. With respect to an NGO observer, EPA provided a favourable recommendation. After considering the precedents cited by Van Putten, he agreed to invite NGO observers to the official US delegation. Similarly, the Canadian Department of External Affairs relied on Environment Canada's judgement.[92] Tom McMillan, Minister of the Environment, wrote to Joe Clark, Secretary of State for External Affairs in August, stating, 'Although I realize that it is unusual to involve the public directly in government-to-government consultative sessions, I believe that the presence of GLU would be useful.'[93]

The Canadian letter of invitation came from Joe Clark, dated September 24, 1987 and addressed to John Jackson, Canadian vice-president of Great Lakes United. In the letter the secretary states:

I am well aware of the work your organization has done in conducting an independent review of the Agreement and of your concern that the Great Lakes be adequately protected from pollution. . . .

We see merit in using the review to try and strengthen the Great Lakes Water Quality Agreement by introducing changes to it in several areas. In keeping with the advice given to governments by the IJC, the Royal Society of Canada/National Research Council of the United States of America and Great Lakes United, we have endeavored to keep key principles and provisions of this basically sound Agreement intact.

With respect to your request for observer status at the bilateral review, you will appreciate that the presence of a binational nongovernmental group at the formal review of an international agreement by its signatories raises some interesting issues of propriety and precedent. Nonetheless, in view of Great Lakes United's credentials as a serious and responsible group and our collective interest in ensuring the best possible review of the Agreement, I am pleased to invite you and one other member of the Canadian section of Great Lakes United to participate as observers to the Canadian delegation.

The Canadian invitation came first and was proffered to John Jackson, asking him to designate one other GLU representative. The GLU board selected Kate Davies, a member of the board of directors and the head of the City of Toronto's Department of Environmental Health. The US State Department's invitation came at the last minute, only nine days before the caucus meeting during which the US negotiating position would be finalized.

It was addressed to David Miller, GLU's executive director, and did not specify who the observer should be. In separate letters the State Department also invited the Sierra Club and the National Wildlife Federation to send observers.

Eder believes that the most important of the events leading to the formal invitation of US observers occurred in Buffalo in August, at one of the public hearings on the agreement organized by the EPA. State Department representatives attended this meeting during which they heard GLU's testimony and heard several others refer to the GLU citizen hearings in their testimony and speak favourably of GLU's leadership.

The letter from the US State Department came from Fred Jones Hall, deputy assistant Secretary of State for European and Canadian Affairs, dated September 30, 1987. Unlike the letter from Canada, it expressed no particular acknowledgement of GLU's efforts or its qualifications for participation. It began, 'I would like to give you an update on the U.S. Government review of the 1978 Great Lakes Water Quality Agreement and invite you to name an observer to our upcoming negotiations with Canada.' It then went into detail about the schedule of upcoming meetings.

Tim Eder, who had been hired as a field coordinator to manage the citizen hearings, was appointed by the GLU Board as GLU's US observer. Mark Van Putten, director of the National Wildlife Federation's Great Lakes Natural Resources Center, represented NWF, and Jane Elder, Great Lakes regional Vice-President, represented the Sierra Club.

Once the Canadian and American delegations were set, including the NGO observers, a dilemma had to be resolved, particularly among the GLU observers. Once they became members of the US national delegation, the observers had to pledge to abide by the requirements for confidentiality. This raised difficult problems for the GLU representatives on each side of the table. For many months before the negotiations GLU members, John Jackson (Canadian) and Tim Eder (American), in particular, had been regularly conferring with each other on strategy and tactics, not only for winning representation at the table but on what positions to advocate once there. Now that each was a member of his country's delegation, this communication and similar discussions with the GLU board and others had to be restricted. This posed a dilemma that none of the observers took lightly. What if either the US or Canadian side tried to weaken the agreement? Would not the GLU representatives feel compelled to confer? Would the required confidentiality contradict one of GLU's fundamental principles, that the interests of the Great Lakes basin superseded narrow national interests? The issue was brought before the GLU board for debate, and the board voted unanimously that the observers should adhere strictly to the confidentiality requirements.

The confidentiality requirement was more troubling philosophically than practically, in this case. In fact, the rest of the technical staffs of the US and Canadian delegations, particularly Ron Shimizu and Kent Fuller, were working so closely together that by the time the separate American and Canadian caucuses met to set their national positions on October 9, the two nations had combined their separate proposals and prepared a unified draft for the caucuses to consider.

Formal caucuses to finalize the national positions were held by the separate US and Canadian delegations, including the NGO observers of each delegation. For the most part the government positions were drafted by the EPA and Environment Canada. Even before being granted observer status, GLU representatives in Canada and the United States had attended a series of meetings with the environmental agency staff during the preparation phase of each country's draft position. So, when the NGOs came to the table, they were thoroughly versed in the proposed amendments and their rationales. The US and Canadian career diplomats played the role of referees, making sure everyone understood and followed the rules of bilateral negotiations. A single formal negotiating session was held between the US and Canadian delegations on October 16, 1987, in Toronto.

Both the Canadian and US NGO observers reported surprise and delight at the role they played. In a summary memo, Tim Eder wrote: 'The entire experience was one I'll remember the rest of my life. It was a tremendous honor for me personally and, I believe, for Great Lakes United as an organization.'

During the formal negotiation the NGOs were present as observers and did not sit at the main table. Much of the work on revisions to the language of the agreement, however, took place in break-out sessions, and there the NGOs were allowed full participation. The NGO representatives were thoroughly involved in discussing every aspect of the agreement. They brought with them a level of technical knowledge and fluency unmatched by any of the other participants, with the exception of the representatives of the environmental agencies. Being part of a binational organization was also a distinct advantage for the GLU observers. They were thoroughly familiar with the proposals from both parties and the internal politics of each and, therefore, had a much deeper understanding of the various proposals than did the other representatives at the table. According to Eder, the NGO representatives were not observers but full participants. The State Department officials recognized them, called on them, and listened to them. As the meetings progressed, the NGO representatives found themselves being turned to with increasing frequency, and few statements were made during which the speaker did not look to them for a nod of approval.

Both Eder and Jackson felt that the process itself, the very active role the NGOs played in the agreement review and amendment, was the most important outcome. According to Eder, 'The fact that we had a seat at the table meant that we took ownership of the Agreement, that we had a stake in it, and as a result we wanted to make sure that it would be implemented.'[94] John Jackson wrote, 'It was widely known that non-government members were included in this stage of the review and renegotiation. As a result, there was a feeling within the environmental community that their concerns were being represented and protected during the actual negotiations.'[95]

In the two years after the signing of the 1987 amendments, public interest in the GLWQA increased immensely. Requests to the IJC's Great Lakes office for agreement-related information rose 162 per cent.[96] This is but one of several continuing expressions of broad public concern in Great Lakes issues.

Although the case described here is unique, it does suggest general propositions for the study of world environmental politics. Nine features of this case and their associated propositions follows.

Point 1

The NGOs involved, particularly Great Lakes United, had embraced the GLWQA and identified closely with its goals and objectives. They had used the terms of the agreement to push both governments to adopt specific Great Lakes policies the NGOs supported. The NGOs had been aware of the scheduled review of the agreement and had anticipated the issues likely to arise. They were particularly concerned about political manoeuvres on the part of the conservative governments of the negotiating parties to weaken the agreement. Their members adopted policy resolutions in defence of the agreement and their leaders made a case for the existence of a public interest in the agreement's future. They pre-empted the governments by holding their own set of citizen hearings on the effectiveness of the agreement, simultaneously raising the political stakes while building a strong case for their formal involvement as citizen representatives.

International agreements and treaties often articulate broad purposes and general goals which are attractive to environmental NGOs. Once adopted, such statements can be used by NGOs as international commitments to support for their issues. Once in the public domain, such agreements become more than simply creatures of the signatory governments. They belong to the class of documents that includes preambles to constitutions and declarations of independence, which articulate a common vision and approach to politics and governance. Their power exceeds simple questions of implementation and enforcement. Even when they are not well implemented – perhaps especially when they are not well implemented – they can be strategically embraced and adopted by NGOs to mobilize public opinion, gain credibility, and pressure parties to adopt policies the NGOs support. When NGOs adopt the defence of such agreements as a tactic, they position themselves as representatives of the public's interest in the agreement's goals.

Point 2

The issues on the table were highly technical and difficult to understand for almost everyone in the negotiations except the environmental agencies and the NGO representatives. The environmentalists, both the government professionals and the NGO representatives, shared a common vocabulary and certain assumptions drawn from their shared understandings of the environmental sciences and their political implications. Career diplomats and bureaucrats from such organizations as the US Office of Management and Budget were likely to defer to those who were fluent in the language of environmental science and regulations. The NGO representatives involved in this case, all highly articulate and knowledgeable, had an influence in the negotiations perhaps disproportionate to what, according to conventional measures, might have been considered their actual political clout. Their particular skill was an ability to translate the language of environment into the language of politics.

The technical complexity of many environmental issues creates a knowledge gap into which NGO representatives can move. Technical expertise and fluency developed by NGOs can lend credibility to their positions and provide important leverage in negotiations. NGOs can translate the technical issues into policy options, articulating their position on the environmental consequences of alternative policy choices.

Point 3

The water quality problems addressed by the 1987 amendments, primarily contamination by persistent toxic chemicals, are, by their nature, diffuse and pervasive throughout the ecosystem. The sources of toxic contamination are multiple and widespread. At the level of individual exposure the effects are subtle, but may aggregate in significance at the level of populations. Like ozone depletion and global warming, specific instances of harm are difficult to find, whereas the credible threats are everywhere. These facts played a subtle role in creating the conditions for NGO participation as citizen representatives rather than representatives of specific stakeholders. The issue of toxics is unlike more traditional water issues in which the stakeholders are limited to users, polluters, and riparians. With the issue of toxic contaminants a stakeholder can be anyone. The traditional concept of stakeholder, therefore, is necessarily broadened by the nature of the pollution as are considerations about who should be represented when stakeholder interests are involved.

By focusing on diffuse environmental issues, such as toxic contamination, global warming, ozone depletion, and loss of biodiversity, NGOs broaden the definition of stakeholder. It is commonly understood that the resolution of these issues lies beyond the capacity of individual governments and, thus, requires international responses. It is equally true, although less well appreciated, that these issues transcend not only national boundaries but traditional 'interest' boundaries, as well. This broadens the range of people with reasons to participate in decision-making and strengthens the argument for why governments ought to listen to a wider range of public voices than those typically involved in resource issues.

Point 4

The NGOs in this case took the principled position that citizen participation was a necessary component of any decision-making regarding Great Lakes water quality. In their own countries they had consistently lobbied for an active role in federal, state, and provincial decision-making. They argued that the public had a compelling interest in the Great Lakes Water Quality Agreement and deserved to have its voice heard wherever decisions about it were being made – even in a forum for formal binational negotiations. The consistency of this position was hard to refute.

The case for NGO involvement in international environmental negotiations is closely linked to the movement for public participation in bureaucratic decision-making. NGOs formulating strategies for their participation in international forums can draw from the extensive experience and literature of public participation, especially in the United States and Canada. The lessons of this experience, however, must be considered in the context of the North American situation and evaluated for applicability in international situations.

Point 5

Although they were invited simply as 'observers', the NGO representatives played an active role both in the preparation of positions in the separate US and Canadian caucuses and in the formal bilateral talks. In addition to their technical fluency the representatives of Great Lakes United had, as a result of the organization's binational character, certain additional advantages over other, more official, participants. Their very presence, with representatives on both sides of the table, affirmed the cross-boundary nature of the issues at hand and challenged the presumption of separate national interests built into the structure of binational negotiations. As the only binational spokespersons present, the status of Great Lakes United's representatives in the negotiations depended on the parties' understanding of the issues as binational. On a more practical level their knowledge of both sides' positions, familiarity with most of the negotiators, and appreciation for inter- and intra-agency politics on both sides of the border gave them an understanding of the issues in greater detail and depth than that of most members of either delegation.

At the negotiating session itself, the protocol followed was for each nation's senior diplomat to welcome the other, after which the floor was turned over to Great Lakes United for comment. Although the government participants understood this as 'good manners and courtesy',[97] this, nonetheless, had the appearance of making the NGOs a third party to the negotiations, equal to the two nation-states.

Transnational NGOs assert by their very existence that environmental issues transcend jurisdictional and bureaucratic boundaries. Their presence, therefore, builds into the structure of the negotiations themselves this transnational, as distinct from bi- or multinational, character. Because of this and the fact that their representatives may have useful information about the internal strategies and resources that the government representatives bring to the table, NGOs may wield power and authority in international negotiations greater than what might be assumed from traditional measures of political power.

Point 6

A true picture of the 'sides' in these negotiations would be even more complex than that of the three sides described. Despite the formal structure of bilateral talks, also represented at the table were the states and provinces, as well as the Army Corps of Engineers, the Office of Management and Budget, Health and Welfare Canada, US Fish and Wildlife Service, Environment Ontario, the NGOs and others. The representatives of the Canadian and US State Departments knew each other well, as did the representatives of their respective environmental agencies, the US EPA and Environment Canada. They shared a common understanding of the issues, and a common vocabulary. Often the professional counterparts across the table – the foreign affairs officers, environmental bureaucrats, along with the NGO members – had more in common with each other than with the members of their respective national delegations. Ron Shimizu[98] has pointed out that before the organized efforts of NGOs in the Great Lakes, an active network of scientists both in and out of government in both countries played a role similar to that played by the NGOs in this case. This network, Shimizu points out, was initiated by the IJC and the governments during the binational water quality studies of the 1960s and 1970s. Scientists felt allegiance not merely to their countries, but to a scientific principle, the ecosystem approach. Kent Fuller[99] noted that the community of scientists and professionals working on shared problems provides transboundary commonality, and even loyalty, that helps the process of international environmental negotiations. The NGOs differed only in that they politicized the scientific principle and adopted self-consciously political strategies for implementing the ecosystem approach.

International environmental negotiations may be carried out by individuals who, although representing different nations, have considerable experience of working with each other and the NGOs and share a degree of common interests. As such, personalities will often play as much, if not more, of a role in international environmental negotiations than the organizations and positions the individuals represent. NGOs will be more effective if they develop common interests and personal relationships with the range of individuals the NGOs are attempting to influence. Common interests can be nurtured and promoted to encourage the development of an international regime of environmental problem-solving.

Point 7

Noticeably absent from the review and amendment process were representatives of industrial or commercial interests. This absence reflects the fact that industry representatives have most often focused their resources and the energy of their environmental personnel on state and provincial regulations and the courts where decisions have immediate financial impact. Since the GLWQA lacked the force of domestic law, industry was not inclined to spend time on questions of definitions, accountability, and goals. Yet industry's absence from the process may prove costly to them. Recent laws passed in the United States refer specifically to the GLWQA and require compliance with its provisions.[100] Hence, by the time industrial lobbyists are fending off new and costly regulation, they find their opponent's cases are strengthened by the weight of international commitments.[101] In the Great Lakes this has proved to be a highly effective environmentalist strategy.

International agreements can reverberate with unforeseen domestic consequences. For NGOs this suggests that gaining influence in international agreements can be a strategy for affecting domestic environmental law and policy. This will vary by nation, the legal standing of the agreement or treaty and the possibilities in each country for influencing the domestic agenda. Success, however, will increase the political stakes and draw to international environmental negotiations the organized interests of polluting industries and their substantial political and economic clout.

Point 8

In the future, particularly if the agreement appears to be driving Great Lakes policies at the state and provincial regulatory level, it is likely that agreement negotiations will receive more attention. Other interest groups, such as industry or associations of local governments and others, may demand representation. The parties will have difficulty limiting representatives. Increased numbers, particularly of interests which do not share the ecological assumptions underlying the agreement, may make the negotiating process unwieldy. The parties may reasonably ask, 'If we invite GLU, why not others? Where do we draw the line?' Hence, some new format, such as the 'round-table discussions' frequently held in Canada, or other public participation mechanisms, may be adopted instead of inviting NGOs officially to the table.

As the stakes rise, more interests will demand access and participation. The formal mechanisms of international negotiations have a limited capacity to accommodate varied interests and their representatives. As a result, new forms of intergovernmental, transnational, and nongovernmental processes will be invented. NGOs should give consideration to proposals for new forums for public participation in international politics.

Point 9

One unintended but predictable result of the 1987 protocols was a pronounced withdrawal by the national governments from the International Joint Commission, an intergovernmental organization (IGO). When preparing for the negotiations, the NGOs appeared to give very little if any consideration to the role of the IJC as an independent international organization with the mandate to monitor progress in achieving agreement ends. In general, the approach toward the IJC by the NGOs has been one of lobbying and politicization of the IJC's role. The NGOs often pushed the IJC to take strong adversarial positions *vis-à-vis* the governments. The IJC often looked to the NGOs to raise difficult political issues the IJC could not and to create a constituency with political clout to help implement IJC recommendations. Both sides of this dynamic saw each other as natural allies. Neither side, however, gave much consideration to how their mandates and objectives might naturally conflict at times. The lack of consideration by the NGOs of the IJC's role may ultimately lead to weakening of the IJC's stature as an effective international organization. This plays into the hands of conservative politicians and bureaucrats eager to reign in the IJC as an authoritative voice for environmental protection and regulation. That voluntary organizations should take over the role of international governance fits neatly into the ideology of *laissez-faire*.

The relationship between NGOs and IGOs is complex and full of pitfalls for all sides. IGOs gain their legitimacy and authority from governments and must work with governments to accomplish their ends. NGOs, however, often work outside government channels – even, in many cases, in opposition to national governments. Yet in world environmental politics considerable mutual interest exists between NGOs and IGOs in creating effective mechanisms for international decision-making. The ideology of voluntarism can be appealing to NGOs. The argument is that nation-states are unable to act effectively in international environmental arenas; therefore, the task of environmental management belongs to the independent sector and voluntary organizations. This strategy is particularly appealing to laissez-faire *liberals concerned about the growing possibility of international regulation of cross-boundary polluting industries and environmentally unsustainable business and trade practices. When developing strategies for participation in world environmental politics, NGOs should develop a clearer understanding of where their interests overlap with IGOs, and consider those overlaps when dealing with governments and corporate sectors.*

One thing is certain, the international politics brought about by ecosystem stresses are here to stay. However one interprets their meanings, or imagines their outcomes, they will remain interesting and likely to affect lives in profound ways.

NOTES

1   The relationships between the federal capitals and their respective states and provinces differ significantly between the US and Canada, particularly with regard to natural resources and pollution control. Canadian provinces have greater control over water resources than the US states. This has required intergovernmental agreement between Canadian federal and provincial powers prior to bilateral or international agreements which impact on provincial resources. For example, the 1971 Canada–Ontario Agreement preceded the US–Canada 1972 Great Lakes Water Quality Agreement. 'The reality is that power under the constitution is in federal hands in one nation [US] and in provincial hands in the other [Canada]' (Carroll, *Environmental Diplomacy* Ann Arbor, MI: The University of Michigan Press, 1983, 30). For a general overview of constitutional and legislative differences regarding pollution control and the Great Lakes, see Carroll (1983) *Environmental Diplomacy*, chapter entitled 'Canadian–U.S. Differences', 28–38 and 130–1. Also see John Carroll, 'Differences in the Environmental Regulatory Climate of Canada and the United States', in *Canadian Water Resource Journal* 4 (1979):16–25. These differences, and others, have affected the development of environmental NGOs in each country. Canadian groups have concentrated their efforts at the provincial and regional levels, while US organizations have concentrated on influencing Washington. Great Lakes United and the Sierra Club organize an annual Great Lakes Week in Washington, but have not done the same in Ottawa.

2   US–Canadian bilateral cooperation regarding the Great Lakes was first formalized by the 1909 Boundary Waters Treaty, which established navigational rights, responsibilities and institutional arrangements for the resolution of disputes over boundary waters. Together, the Boundary Waters Treaty and the International Joint Commission, a quasi-supranational binational commission established by the treaty, have evolved to provide a framework for bilateral environmental cooperation between the US and Canada regarding the Great Lakes.

3   International Joint Commission, *Second Biennial Report*, Windsor, Ontario: International Joint Commission, December 31, 1984, 10).

4   These concepts have evolved over the life of the Agreement. For example, the 1972 GLWQA area of purview – boundary waters – was largely delimited by national jurisdictional boundaries, rather than by ecological parameters. However, research undertaken following the 1972 GLWQA established the basis for the ecosystem approach which was embodied in the Great Lakes Water Quality Agreement of 1978. See below for further discussion of this evolution.

5   Many of the world's largest lake basins and over 200 river basins cross international boundaries. Notable international commissions for water resource management include the International Commissions for the Protection of the Rhine, Niger River, Lake Chad basin, and the Mekong River basin. Naturally, each is differentiated by its political and ecological context. For a comparative examination of the institutional arrangement in the Great Lakes and other international basins see Dante A. Caponera, 'Patterns of Cooperation in International Water Law: Principles and Institutions', also Teclaff and Teclaff, 'Transboundary Toxic Pollution and the Drainage Basin Concept', both in *Transboundary Resources Law*, Albert Utton and Ludwik A. Teclaff, eds, (Boulder, Co: Westview Press, 1987). Also see Lynton K. Caldwell, 'Regional Arrangements: Bilateral and Multicultural Agreements' in *International Environmental Policy*, Durham, NC: Duke University Press, 1990.

6   The IJC Science Advisory Board, *1989 Report to the International Joint Commission*, declared that 'the significance of the 1978 Great Lakes Water Quality Agreement lies in its strong affirmation of the need for an integrated

ecosystemic social-economic-environmental approach to problem solving.' It went on to suggest that cooperative institutional arrangements which have evolved to implement GLWQA may serve as a model for international cooperation on biospheric problems. On a more regional level, institutional arrangements in the Great Lakes have been widely noted as model basin-wide water resource management (Caponera, 1987).

7 This governance structure may be referred to as an international regime for natural resources, as referred to by Oran R. Young who defines a regime as 'social institutions governing the actions of those involved in specifiable activities or sets of activities. Like all social institutions, they are practices consisting of recognized roles linked together by clusters or conventions governing relations among the occupants of these roles' (Young, 1989:12). An excellent conceptual analysis of international resource regimes as well as supporting case studies can be found in Oran C. Young's *International Cooperation: Building Regimes for Natural Resources and the Environment* Ithaca, NY: Cornell University Press, 1989.

8 By which I mean that informal system of professionals with environmental expertise who serve governments, universities, NGOs, international agencies and who share a common vocabulary and international scientific culture. Haas has termed this an 'epistemic community'. See Peter Haas, *Saving the Mediterranean: The Politics of International Environmental Cooperation*, New York: Columbia University Press, 1990: 52–63. For a thorough treatment of the role of science and epistemic communities in resource management, see articles in Andersen and Ostreng, eds, *International Resource Management: The Role of Science and Politics*, London: Belhaven Press, 1989.

9 International Joint Commission, Science Advisory Board. *1989 Report to the International Joint Commission* Windsor, Ontario: International Joint Commission, 1989: 5.

10 Michael Donahue, personal communication 1991.

11 Kent Fuller, interview 1991. Lynton Caldwell has described the increased emphasis on the states, provinces and national governments:

> From one viewpoint, this might appear to be a positive response; but from another it could be seen as counteracting regional initiatives by strengthening the hands of national agencies that hitherto had been less than vigorous in leading toward water quality and basin-wide ecosystem objectives.

Lynton Caldwell, 'Emerging Boundary Environmental Challenges', *Natural Resources Journal*, 33 (1), Winter 1993.

12 At the turn of the century bilateral disputes surrounded development of the St Lawrence Seaway for navigation, unilateral construction of the Chicago Diversion Canal, the St Mary River–Milk Irrigation Project and the hydroelectric project, as well as proposals for the St Mary River at Sault Ste Marie in Michigan and Ontario. (See Carroll, *Environmental Diplomacy*, 40.)

13 This institutional arrangement has had considerable influence on the capacity of the IJC to implement GLWQA and gain compliance on the terms of the agreement. It has been a recurring issue for the IJC and proponents of a more centralized supranational authority. This limitation on authoritative power also differentiates the IJC and, thus, the Great Lakes ecosystem approach from the basin-wide management approaches employed at other international basins, such as that of the River Rhine. See Caponera, 'Patterns of Cooperation'.

14 The first real attempt to study Great Lakes water quality was initiated by the Mayor of Chicago in 1908, when he organized an interstate effort to study the pollution of Lake Michigan. Later that year several Lake Erie cities did the same for Lake Erie.

15 The International Joint Commission 1970 Report (Ottawa, Ontario: International

Joint Commission, 1970) which provided impetus for the 1972 GLWQA, cited pollution in the lower Great Lakes that was in 'contravention of the 1909 Boundary Waters Treaty'. See National Research Council of the United States and the Royal Society of Canada, *The Great Lakes Water Quality Agreement: An Evolving Instrument for Ecosystem Management*, (Washington, DC: National Academy Press, 1985), 20–3 for an overview of IJC-referenced studies before the 1972 Great Lakes Water Quality Agreement.

16 Quoted in Leonard B. Dworsky and Charles F. Swezey, *The Great Lakes of the United States and Canada: A Reader on Management Improvement Strategies* (Ithaca, NY: Cornell University Press, 1974, 38).

17 University of Wisconsin Sea Grant Institute, *The Fisheries of the Great Lakes*. Report, Board of Regents, University of Wisconsin System Sea Grant Institute, 1988.

18 Personal, scientific, and newspaper accounts of ecological degradation resulting from pollution are well documented in Phil Weller, *Fresh Water Seas: Saving the Great Lakes* (Toronto: Between the Lines, 1990), 89–93. For various accounts of ecological degradation in the Great Lakes, see William Ashworth, *The Late Great Lakes: An Environmental History* (New York: Alfred A. Knopf, 1986), 123–48.

19 The Water Quality Board is composed of ten American and ten Canadian managers of pollution control programmes appointed by the commissioners on recommendation from the government agencies. Appointees to the Science Advisory Board (SAB) were drawn largely from the community of experts who were involved since the 1960s in studies on Great Lakes pollution questions.

20 A new addition to IJC structure has been the Council of Great Lakes Research Managers, originally set up to serve the SAB as a direct connection to those responsible for prioritizing and funding research on Great Lakes problems. In the most recent reorganization of the IJC committee structure the council was raised in status to equivalence with the Water Quality Board (WQB) and SAB and now reports to the commission.

21 The Canadian–US Council of Great Lakes Research Managers recently reported on the Great Lakes research community's contributions, research priorities, and future challenges in the Council of Great Lakes Research Managers Futures Workshop (Ottawa: Rawson Academy of Aquatic Sciences, 1989). For a summary of the workshop see 'Great Lakes 2000: Building a Vision', International Joint Commission, 1989. For an in-depth treatment of the role of science in the formation and transformation of international regimes or governance structures, see Andersen and Ostreng, *International Resource Management*.

22 The first session of CUSIS was held from December 1971 to June 1972 and focused on strengthening institutional arrangements for international resource management in the Great Lakes. See Dworsky and Swezey, *Great Lakes*. See also Leonard B. Dworsky, 'The Great Lakes: 1955–1985', in Lynton K. Caldwell, *Perspectives on Ecosystem Management for the Great Lakes: A Reader* (Albany, NY: SUNY Press, 1988).

23 It is useful to note that the Boundary Waters Treaty contained a mandate for public hearings in conjunction with applications for approval for engineering works. The IJC had also always held hearings before the issuance of their recommendations to the governments. These were part of a formal process, and the public tended to be intimidated, particularly in light of the increasingly technical nature of the reports on which the public was being asked to comment. Interview with Mimi Becker, past president, Great Lakes Tomorrow, in Ann Arbor, MI (Oct. 17, 1991) ('Becker interview').

24 Becker interview, *supra* note 26.

25 Robert J. Mason, 'Public Concerns and PLUARG: Selected Findings and Discussion', *Journal of Great Lakes Research*, 6(3): 210–22.

26 A. P. Grima and R. J. Mason. 'Apples and Oranges: Toward a Critique of Public Participation in Great Lakes Decisions', *Canadian Water Resources Journal*, 8(1): 22–50.

27 As quoted in Grima and Mason, 'Apples and Oranges', 40.

28 Grima and Mason note that a 'very limited' public was involved in the PLUARG consultation panels, and the 'general public remained generally unaware of the PLUARG study.' They, however, further note that 'the panels were representative of most potentially affected interests' (Grima and Mason, 'Apples and Oranges', 40).

29 Mason, 'Public Concerns and PLUARG', 210.

30 Grima and Mason, 'Apples and Oranges'; Carol Y. Swinehart, 'A Review of Public Participation in the Great Lakes Water Quality Agreement', in Hickcox, David H., ed., *Great Lakes: Living with North America's Inland Waters* (American Water Resources Association, Symposium Proceedings, November 1988).

31 The *Great Lakes Communicator*, 9(1): 7, October, 1978. The *Communicator* is a monthly newsletter published by the now defunct US Great Lakes Basin Commission, a state-federal water resource planning agency established under the Water Resources Planning Act of 1965.

32 Swinehart, 'Review of Public Participation'.

33 Becker interview, *supra* note 26.

34 'PLUARG's more than one hundred reports [during its six-year study] were crucial in putting the ecosystem approach at the core of the 1978 agreement', Theodora E. Colborn *et al.*, *Great Lakes, Great Legacy?*, (Washington, DC: The Conservation Foundation and Institute for Research on Public Policy, 1990). Caldwell also notes the significance of PLUARG in incorporating the ecosystem approach in the 1978 GLWQA (Caldwell, *Perspectives on Ecosystem Management*, 31).

35 Ohio was the exception – it did not enact a detergent phosphorus ban until 1989.

36 Weller, *Fresh Water Seas*. For more detailed accounts of scientific findings regarding the effect of toxics on Great Lakes biota, see Michael Gilbertson, 'Epidemics in Birds and Mammals Caused by Chemicals in the Great Lakes', in *Toxic Contaminants and Ecosystem Health: A Great Lakes Focus*, ed. Marlene S. Evans (New York: John Wiley, 1988). Colborn, *Great Lakes: Great Legacy?*, 113–85, thoroughly covers ecological and human health impacts of toxics in the Great Lakes, supported by an extensive list of references.

37 For an overview of bioaccumulation of toxics and related impacts on Great Lakes species see R. J. Allen, *et al.*, *Environment Canada, Toxic Chemicals in the Great Lakes and Associated Effects* (1991) the synopsis or full-length Environment Canada report, *Toxic Chemicals in the Great Lakes and Associated Effects* (Environment Canada: Government of Canada, Ottawa, 1991).

38 The 1978 GLWQA includes Lake Michigan, which was not previously considered part of the 'boundary waters' because it is geographically situated entirely within US jurisdiction. Its inclusion in the 1978 GLWQA is one example of the increasing predominance of an ecosystem perspective and consideration of ecological factors. Commenting on further shifts in this direction represented by the 1987 protocols, the Council of Great Lakes Research Managers summarized the two-decade transition (Futures Workshop, Ottawa: Rawson Academy of Aquatic Sciences, 1989):

> The transition of the Agreement from an Agreement on water quality [in 1972] to an Agreement on water quality in an ecosystem context [1978], to an Agreement on managing the human uses and abuses of the Great Lakes Basin Ecosystem, is not completed but the 1987 protocols represent significant movement in this direction. (121)

39 International Joint Commission, United States and Canada. Revised Great Lakes

Water Quality Agreement of 1978 (International Joint Commission, September, 1989).

40 For a treatment of these concepts from an advocacy perspective, see Program for Zero Discharge, 'A Prescription for Healthy Great Lakes' (Washington, DC: National Wildlife Federation and Canadian Institute for Environmental Law and Policy, 1991).

41 The problems and contradictions inherent in translating 'an ecosystem approach' into meaningful action have been fairly well explored, although not yet resolved. See Lynton K. Caldwell, ed., *Perspectives on Ecosystem Management for the Great Lakes* (Albany, NY: SUNY Press, 1988). Caldwell's introductory chapter offers a general explanation of terms of reference and issues. See also other articles in the same text, which examine limitations of institutional arrangements, particularly D. Munton 'Toward a More Accountable Process: The Royal Society-National Research Council Report'. A 1985 Ecosystem Approach Workshop, held in Hiram, Ohio, explored obstacles to implementation. See summary report, W. J. Christie, *et al.*, 'Managing the Great Lakes Basin as a Home', *Journal of Great Lakes Research*, vol 12: 2–17. Barry Boyer offers a more legalistic analysis in 'Ecosystem, Legal System and the Great Lakes Water Quality Agreement' (presented at the annual meeting of the International Association of Great Lakes Researchers (IAGLR), Buffalo, NY, June 1991). Constitutional limitations are explored by Jack Manno in 'Federalist and Ecologist' (typescript, 1991).

42 International Joint Commission, *Second Biennial Report*, 5.

43 Tim Eder, interview, 1991.

44 Oran R. Young has also observed this to be true of NGOs and the Great Lakes Water Quality Agreement as well as within other international regimes:

> International regimes also commonly give rise to nongovernmental interest groups committed to defending the provisions of specific regimes and prepared to press governments to comply with their own dictates. In fact, the establishment of a regime can stimulate the growth of powerful interest groups in a number of the member states which then form transnational alliances in order to persuade responsible agencies to comply with the requirements of the regime (i.e., Mediterranean Action Plan and GLWQA 1972, 1978).
>
> *(International Cooperation, 78)*

45 Sierra Club brochure.

46 Sierra Club, *Great Lakes Washington Report* 1(1).

47 Weller, *'Fresh Water Seas'*, quotes Governor James Blanchard who described GLU as 'informed, effective and influential'. The Institute for Research and Public Policy and the Conservation Foundation together noted GLU's prominent role in binational cooperation, Colborn *et al.*, *Great Lakes, Great Legacy?*, 217. See also Sally Lerner, 'A Study of Ontario Volunteer Environmental Stewardship Groups', Technical Paper no.6, Heritage Resource Center, Waterloo, IA, 1991.

48 See William Leiss, ed., *Ecology vs. Politics in Canada* (Toronto: University of Toronto Press, 1979).

49 Ronald J. Engel, *Sacred Dunes. The Struggle for Community in the Indiana Dunes* (Middletown, CT: Wesleyan University Press, 1983).

50 Weller, *Fresh Water Seas*.

51 Steven Schatzow, 'The Influence of the Public in Federal Environmental Decision-making in Canada' in Derrick Sewell and J. T. Coppock, *Public Participation in Planning* (New York: John Wiley, 1977, 148).

52 For an informative overview of the origins and development of the North American environmental movement, particularly in regard to toxic pollution, see Robert C. Paehlke, 'Conservation, Ecology and Pollution', in *Environmentalism*

and the Future of Progressive Politics (New Haven, CT: Yale University Press, 1989).

53 For a valuable discussion of the concept of regulatory culture, see Errol Meidinger, 'Community, Culture and Democracy in Administrative Regulation' (Working paper of the Baldy Center for Law and Social Policy, Buffalo, NY, 1991).

54 I have briefly addressed this issue in 'Citizen Participation and Consensus Building', in *Environmental Dispute Resolution in the Great Lakes Region*, eds, L. S. Bankert and R. W. Flint (Buffalo, NY: Great Lakes Program, Great Lakes monograph no. 1, November. 1988). See also John Carroll's chapter on Canadian–US differences in *Environmental Diplomacy.*

55 'Great Lakes Coalition Proposed By Group', *Watertown Daily Times*, December 1, 1981, at 1.

56 There are several good analyses of the positions concerned that characterized the conservation movement. One I find useful is Samuel P. Hays, *Conservation and the Gospel of Efficiency: The Progressive Conservative Movement 1890–1920*, Harvard Historical Monographs (Cambridge, MA: Harvard University Press, 1959).

57 The implications of ecosystem politics carry many intrinsic difficulties. Not the least of these is the simultaneous expansion in the understanding of the significance of both the impacts of local processes on the broad ecosystem and the impact of broad ecosystem processes on local environments. Politically, advocates are required not only to think globally and act locally, but to think locally and act globally as well. This has posed some difficult challenges, in particular in the RAP programme as described later. This phenomenon has not been well understood. The difficulties it creates have often come as a surprise to those involved.

58 'Lakes Topic of Meeting', *Mining Journal*, Marquette, MI, November 21, 1981.

59 David Everett, 'Great Lakes Coalition Born', *Detroit Free Press*, November 22, 1982.

60 'Abbie Hoffman, MUCC Split With Great Lakes Group', *The Bay City Times*, November 22, 1982, at 10A

61 'Great Lakes Coalition Born', *supra* note 100.

62 Although the level of public participation and type of input differed, in both instances participation was sought basin-wide from both Canada and the United States to gain public support and credibility. The levels and types of participation differed somewhat in the sectors of public involved and final documentation. Because each was generally successful in attaining its objectives, they greatly influenced subsequent bilateral negotiations.

63 Compare with PLUARG stated objectives in this chapter.

64 Great Lakes United, Water Quality Task Force, *Unfulfilled Promises: A Citizen's Review of the International Great Lakes Water Quality Agreement* (Buffalo, NY: Great Lakes United, 1987).

65 Ibid.

66 Ibid.

67 The RSCNRC Report (1985) is a major binational collaborative assessment, by the leading non-governmental scientific organizations, of the ecosystem approach as committed to under the GLWQA. Don Munton summarizes the recommendations and conclusions of the report. Regarding Great Lakes institutions, he concludes that an ecosystem approach can be achieved through changes in the processes of governance rather than the structures of governance, thus requiring greater accountability on commitments made under GLWQA to the public. The report, he notes, calls for 'responsibilities and commitments of the governments to be made clear and unambiguous'. The report ultimately recommends that the

IJC not be involved in the coordination of government programmes. See Munton, 'Toward a More Accountable Process'.

68 National Research Council of the United States and The Royal Society of Canada, *Great Lakes Water Quality Agreement*.

69 Ron Shimizu, personal interview, 1991.

70 Ibid.

71 Kent Fuller, personal interview, 1991.

72 For a current overview and assessment, see International Joint Commission, Great Lakes Water Quality Board, *Review and Evaluation of the Great Lakes Remedial Action Plan (RAP) Program* (Windsor, Ontario: International Joint Commission, 1991).

73 Carroll, *Environmental Diplomacy*.

74 International Joint Commission, 'Remedial Action Plans for Areas of Concern', Informational brochure (Windsor, Ontario: IJC, 1985).

75 See Swinehart, *A Review of Public Participation*, in which she reviews participation in IJC public hearings.

76 This trend continued beyond the period discussed here. The 1989 biennial meeting of the IJC was dominated by Great Lakes activists. More than 800 people attended, many of them brought in for the event by Greenpeace, which had recently organized a Great Lakes international office. Greenpeace had promised to hold a separate meeting at the time of the IJC biennial to protest and, perhaps, disrupt the official meetings. In a move to pre-empt that scenario, Greenpeace's spokesperson, Joyce McLean, was invited by the IJC to give the keynote address. Her speech was preceded by a ritual procession of puppets and banners representing the fish and wildlife harmed by toxic pollution. The contrasts between the countercultural rite and the pomp and circumstance of an official international gathering, replete with guards of honour and plenty of flags was strikingly worthy of the legacy of Abby Hoffman. From the podium Joyce McLean chided the commissioners on how rarely their recommendations had been heeded by the parties. Perhaps, she suggested, they should resign? Instead, they issued a report on the 1989 meeting in which they said, in addition to the comment that heads this study, 'The increasing level of public concern for the Great Lakes ecosystem and insistence on governmental response to Agreement objectives were strikingly evident and outspokenly vented, at the Commission's recent Biennial meeting' (IJC, *Biennial Report*, 5). The fact that Joyce McLean's comments were memorable could be seen at the 1991 biennial when the US chairman of the IJC, Gordon Durnil, referred to them several times as being responsible for his personal reconsideration of his role as a commissioner. The 1991 biennial was also noteworthy for the number of times the commissioners reminded the public that the IJC's role was merely advisory and that the commissioners had no power to implement the programmes some activists were urging. Ironically, Joyce McLean attended the 1991 biennial as a newly appointed official in the environment ministry of the Ontario Provincial government.

77 International Joint Commission, *First Biennial Report under the Great Lakes Water Quality Agreement of 1978* (Windsor, Ontario: International Joint Commission, June 1982).

78 US State Department, *US Response To the International Joint Commission's First Biennial Report Under the Great Lakes Water Quality Agreement of 1978* (February 1978). Quoted in Joseph T. Jockel and Alan M. Schwartz, 'The Changing Environmental Role of the Canada-United States International Joint Commission', *Environmental Review* vol. 8, 1984, 236, 247.

79 NRCRSC, *Great Lakes Water Quality Agreement*, 5

80 Kent Fuller, personal communication, 1991.

81 For example, in draft amendments to the GLWQA prepared by the US EPA, the

responsibility for carrying out citizen participation programmes in RAPs would be given to the IJC. Despite the fact that GLU had frequently encouraged the IJC to undertake public participation programmes, GLU responded that it believed

> it is inappropriate to put the onus of responsibility for public information and involvement in RAPs on the Great Lakes Regional office of the IJC as required on page 13 [of the EPA Proposed Changes All-Annex Format. Public, unreleased document. EPA, 1987]. We find it hard to believe that the states and provinces would allow the Regional Office to take such an active role in plan preparation. Further, the primary responsibility for preparing RAPs lies with the states and provinces. Because public involvement must be integrated throughout the RAP preparation process, the states and provinces bear the responsibility for public involvement. A more appropriate role for the Regional Office would be to prepare public involvement guidelines.

GLU Files, Buffalo, NY.

82 This is strictly true. Elizabeth Dowdswell, Environment Canada, and Val Adamkus, US EPA, however, both represented their agencies in the review and negotiation process, and at the time they were Canadian and American co-chairs, respectively, of the IJC Water Quality Board.
83 The IJC's position has eroded since the signing of the 1987 amendments. The United States, in particular, appears to have withdrawn some support from the IJC, as can be seen in the limited participation of senior officials from federal and state agencies in the RAPs. IJC officials have had increased difficulty obtaining necessary information from US government agencies relative to monitoring the GLWQA objectives. At the same time, and as a result of the 1987 protocols shifting reporting responsibilities to the governments, the IJC dissolved its committee structure, in which governmental officials had traditionally contributed their support to IJC activities.
84 Swinehart, in 'Review of Public Participation'.
85 EPA Agreement Summary, public, unreleased document. Available in two forms: All Annex Format Draft and Index to Proposed Changes, Original Draft Form. This document is a draft of proposed changes to the Great Lakes Agreement (original format). Published by EPA, 1987.
86 Mark Van Putten, personal interview, 1991.
87 Tim Eder, personal interview, 1991.
88 Ibid.
89 Ibid.
90 John Jackson, Tim Eder, Ron Shimizu, personal interviews, 1991.
91 Kent Fuller, personal interview, 1991.
92 Ron Shimizu, personal interview, 1991.
93 John Jackson, 'Citizen Involvement in the Review of the Great Lakes Water Quality Agreement', unpublished paper, 1991. Available through the Great Lakes Research Consortium, 24 Bray Hall, SUNY ESF, Syracuse, NY, 13210.
94 Tim Eder, personal interview, 1991.
95 Jackson, 'Citizen Involvement'.
96 John Jackson and Tim Eder. 'The Public's Role in Lake Management: The Experience in the Great Lakes', in International Lake Environment Committee *Socioeconomic Aspects of Lake/Reservoir Management*, (Shinga, Japan: United Nations Environment Programme, 1991).
97 Ron Shimizu, personal communication, 1991.
98 Ibid.
99 Kent Fuller, personal communication, 1991.
100 See, e.g., The Great Lakes Critical Programs Act of 1990, Pub. L. No. 101–596, section 1, 104 Stat. 3000 (codified at 33 USC section 1268 [Supp.1990]).

101 Industry evidently learned this lesson before the 1993 biennial meetings of the International Joint Commission. There, several hundred industry representatives participated in debates from the floor, in public information sessions scheduled in conjunction with the IJC meetings, and in the press. The focus of their attention was an earlier Commission recommendation to the US and Canadian governments that they begin a process leading to phase-out of the use of chlorine as an industrial feedstock in Great Lakes industries.

5 The ivory trade ban: NGOs and international conservation

Thomas Princen

Five hundred years ago, an estimated 10 million elephants populated Africa's forests and savannas, from the shores of the Mediterranean almost to the tip of the Cape of Good Hope.[1] Although figures vary, the elephant population in the late 1980s was believed to be one-twentieth of that figure. In just one decade, the elephant population dropped nearly 50 per cent, from an estimated 1.3 million in 1979 to 625,000 in 1989.[2] At this rate, many observers predicted the elephant would be extinct in many parts of Africa by the end of the twentieth century. Concerned about such a threat to a popular, keystone species, and convinced that trade and consumption patterns contributed to its decline, the international community responded in the late 1980s with a quota system and then a global ban on the ivory trade.

Environmental NGOs were key players in this process, but, as will be seen, their success in getting the ban had ambiguous effects, in part because blunt international instruments like quota systems and trade bans can never address local resource and social conditions. Nevertheless, their efforts do constitute a set of transnational, environmentally oriented linkages parallel to transnational, commercially oriented linkages. The case thus provides an example of how NGO relations constitute a form of resistance to dominant economic relations, as discussed in Chapter 2. It also illustrates how social learning can occur at the organizational, societal, and international levels, as discussed in Chapter 3.

The purpose of this chapter is to show how, when intergovernmental organizations and national governments are unable to stem the decline of a traded species, environmental NGOs can step in to play a critical role. I argue, first, that NGOs promote international actions such as bans because other actors are reluctant or unable to address imminent irreversibilities and second, that NGOs attempt to link local resource conditions in range states to global economic conditions because high demand and fluid trade patterns make resource exploitation unsustainable. I begin by setting the stage with the biophysical and social conditions surrounding the decline in elephant populations and the international responses. I then assess bans as an environmental tactic, the wildlife trade regime (with heavy NGO participation) as a biodiversity regime, and trends in international conservation and in NGO roles.

ELEPHANTS, IVORY, AND INTERNATIONAL CONSERVATION

The continent-wide decline in the African elephant was reflected in the decline of many, but not all, individual populations. Kenya's elephant population dropped from 65,000 in 1980 to 18,000 in 1990, even though all ivory trade had been outlawed in Kenya since 1978. Between 1979 and 1987 the elephant population in Tanzania's Selous Game Reserve, one of Africa's largest, plummeted from nearly 316,000 to 85,000.[3] Uganda has lost 85 per cent of its elephants since 1973. Tens of thousands of illegally obtained elephant tusks have been smuggled out of west and central African nations such as Tanzania, Zambia, Zaire, Sudan, and Somalia.[4] At the same time, however, populations in southern Africa, especially those in Botswana, Zimbabwe, and South Africa, have stabilized or even increased as a result of effective ranch management practices. Despite the apparent successes in southern Africa, legal trade conducted by these countries threatened populations elsewhere. Through the 1980s evidence mounted that the southern African countries were being used for illegal transshipment of ivory from the north to foreign destinations.

Ecologists attribute the drastic decline in African elephants to several factors. Severe droughts, hunting for meat or trophies and, although rare, natural death accounts for some of the decline. But as with threatened species worldwide, loss of habitat to human encroachment poses the most serious long-term threat to the elephant's survival.[5] When Africa had ten million elephants, only sixteen million people lived on the entire continent. Today, Africa's human population has reached five hundred million, and the elephant's range has been reduced to less than one-fourth of the continent's surface.[6] The tropical and subtropical realms where elephants dwell are precisely where the human population in Africa has been increasing fastest, quadrupling in number since the turn of the century, claiming more and more elephant range for cropland, timber, and pastureland.[7] Elephants and cattle compete for some of the same food. With Africa's human population projected to double in twenty-four years, elephant habitat will dwindle rapidly.

Although habitat loss is the greatest long-term threat, the principal immediate cause of elephant deaths in recent years has been poaching. After World War II, the demand and price for ivory rose sharply, making poaching a very lucrative business. In the 1970s, many people invested in ivory as a hedge on worldwide inflation, and ivory prices quadrupled. Prices for raw ivory in Japan, Hong Kong, and Europe, the major ivory-consuming countries, rose from between $3 and $10 a pound in the 1960s to $50 a pound by the mid-1970s. In the 1980s, ivory sold for as much as $200 a pound. At the same, time Hong Kong vastly improved its ability to mass produce ivory carvings.[8] Consequently, international trade in ivory and, thus, elephant killing, sky-rocketed. By one estimate, poachers were killing 200 to 300 elephants a day.[9]

With poaching on the rise, African governments attempted to combat it, but many lacked the resources. Growing political turbulence and easy access to automatic weapons greatly reduced producer governments' ability to protect elephant herds and enforce game laws. Paramilitaristic poachers who used methods such as spraying bullets from semi-automatic weapons over entire herds to collect the ivory of a few elephants could just as easily turn their weapons on ill-equipped, poorly trained game wardens. In some cases, political corruption undermined conservation and protection efforts. Moreover, poaching and trade sometimes flourished as a by-product of regional or civil wars. During the Rhodesian war, for example, Rhodesian special forces formed their own ivory smuggling rings, becoming increasingly active as the tide turned against the government.

Poaching not only reduces elephant numbers but threatens the viability of herds by disrupting the elephants' social structure. As populations dwindle, poachers kill increasingly younger adult elephants. Although elephants can live for up to sixty years, some regions have reported that no elephants more than thirty years of age can be found.[10] Because elephants acquire survival skills over a lifetime and pass their knowledge on to younger elephants, this knowledge is lost as more and more older elephants are killed. Increasingly younger herds may not be as capable of raising and protecting their young.

Elephants are a major force in maintaining biological diversity in the African savanna and forests. As elephants browse in woody vegetation, debarking and pushing over trees and saplings, they disperse seeds and block bush invasion in savannas and woodlands, thereby creating a more productive flora for grazing ungulates and grassland species. Likewise, elephants play a major role in the dynamic savanna–woodland balance. After the elephants leave a savanna, it grows into scrub for a host of browsing animals and then, once more, becomes woodland, to which elephants will return and repeat the cycle.[11] Many forest animals prosper from the elephants' presence. Mongooses, velvet monkeys, and baboons feed on seeds and insects found in elephant droppings. Beetles roll and bury balls of elephant dung as a food supply for their larvae, which honey badgers later dig up and feast upon.[12] Commercial livestock economies also benefit from healthier herds as elephants expand grasslands and reduce the incidence of the tsetse fly.

Increasing ivory prices can be largely attributed to rapidly increasing consumer income and to changes in consumer preferences, especially in East Asia. The Japanese have used ivory in jewellery, trinkets, ornaments, personal signature seals (*hanko*) which are required on official documents, and in medical, fertility, and religious rites. The popularity of signature seals, especially, grew along with Japan's affluence until the seals accounted for the single largest use of tusks anywhere.[13] Europeans have used ivory principally for decoration, as well as for piano keys, knife handles, and musical instruments such as bagpipes. Ivory jewellery, billiard balls, and art have been popular in the United States.[14]

Through the 1970s and 1980s, Japan and Hong Kong were the world's leading ivory-consuming countries. Many tusks passed through outposts in

Dubai, United Arab Emirates (UAE), and Singapore. In 1984 alone, Japan imported some $30 million-worth of raw and worked ivory from Africa. An estimated 80 per cent was illegal. Japan is reported to have accepted shipments of poached ivory from Zaire, Sudan, and the Congo, where ivory exports were banned, from Burundi, which has no native elephants, and from Uganda, where export permits had been forged.[15] Japan was the ultimate destination for 40 per cent of all the world's ivory.[16]

Hong Kong used ivory principally for carving. Some 75 per cent of the ivory carved in Hong Kong in the early 1980s was exported, and much of the rest was sold to foreign tourists. Some 40–50 per cent of Hong Kong's total exports went to the United States.[17] US consumer demand thus accounted for roughly fifteen of every one hundred elephants killed.[18] Worked ivory from Hong Kong was also exported to such countries as France, the Federal Republic of Germany, Italy, Austria, the United Kingdom and Japan. The European Community accounted for roughly twenty of every one hundred elephants killed.[19] In 1984, after the United Kingdom implemented stricter trade controls in Hong Kong, carvers effectively exported their expertise, setting up offshore factories in Macao, Singapore, Taiwan, and the UAE.[20] After ivory was worked in a carving country, it was nearly impossible to detect its origin or determine whether the ivory had been obtained legally or illegally. Either way, the carved ivory easily entered the international wholesale and retail markets.

Although consumer demand in Europe and the United States accounted for a significant portion of the ivory trade, the most important factor contributing to the growth in ivory demand was income growth in East Asia, especially in Japan.[21] Ivory has been a culturally valued substance in Asia for many centuries, but Japan's newfound wealth led to rapidly increased consumption. Japanese incomes have increased faster and more constantly than any others in the world. And the Japanese taste for ivory increased even faster. According to one study, from the mid-1960s to the late 1980s, for every doubling of Japanese incomes, net imports of raw ivory increased by 150 per cent or more. Thus, between 1960 and 1985, as world consumption of unworked ivory increased by 100 per cent, Japanese consumption increased by 200 per cent.[22]

From an economic perspective, it is doubtful whether the ivory trade substantially benefited most producer countries, although the southern African countries' ivory exports did support wildlife management programmes and local communities.[23] Many African producer states argued that export earnings from the ivory trade were badly needed to support their struggling economies. Yet revenues obtained from ivory between 1979 and 1987 amounted to less than 2 per cent of total merchandise export earnings for all but seven African states. Only the Central African Republic and the Congo earned significant revenues (more than 5 per cent of earnings) from trading ivory. Other than Zimbabwe, most African nations sold their ivory at only 10 to 20 per cent of the prevailing selling price in Hong Kong.

Consequently, Africa received a fraction ($10–$20 million) of the $50 million gross ivory revenue per annum. Some of this revenue was lost to shipping costs, but approximately $30–$40 million was captured by Asian exporters and stockpilers. As a result, the actual value of the ivory trade to Africa was not as great as it appeared. The benefits were primarily garnered by non-Africans.[24]

Not only did producer countries gain little from the ivory trade, they may even have lost revenues as tourism declined. Kenya alone has 250,000 to 300,000 tourists annually who indicate an interest in viewing elephants. In one survey, tourists indicated they would travel elsewhere if Kenya experienced a continued elephant population decline. If tourism did decline, by one estimate Kenya could lose between $50 and $80 million annually – up to one-half its total tourist industry revenue.[25]

The first international response to declining elephant populations occurred when Ghana listed the African elephant on appendix III of the Convention on International Trade in Endangered Species of Wild Fauna and Flora (CITES) in 1976. This listing, however, afforded the elephant minimal protection.

As populations continued to decline, the CITES Conference of the Parties of 1978 uplisted the species to CITES appendix II, conferring additional protection. But populations continued to drop, whereupon, in 1985, at the request of several African countries, the parties established export quotas and a trade control system.

The Ivory Trade Control System went into effect in 1986 and was overseen by the CITES Secretariat and managed by a special Ivory Control Unit. The system called for each African state wishing to export raw ivory to communicate to the secretariat its intended export volume, expressed in maximum number of tusks, for the following year. These volumes constituted the 'quotas'. The secretariat had no mandate to unilaterally alter or reject the volume submitted by sovereign states. The secretariat also had to rely on producer states to ensure that the quotas did indeed help conserve elephant populations. The system's effectiveness over a four-year period was the subject of considerable debate among governmental, intergovernmental, and non-governmental organizations.

According to the Wildlife Trade Monitoring Unit (WTMU), a London-based organization contracted by CITES, the total trade in raw ivory fluctuated between 600 and 1,160 tons per year from 1979 to 1986, declined sharply to 370 tons in 1987 and then dropped to 153 tons in 1988. The CITES Secretariat argued that the reduction in ivory trade volume, the drop in demand for ivory, and the decline in carving industry revenue could be linked to the ivory quota system.[26] Although the decline occurred subsequent to the introduction of the control system in 1986, WTMU argued that, at the time, it was too early to draw firm conclusions about the true impact of the CITES ivory export system.

The CITES Secretariat also argued that, despite some early problems, procedural changes made in 1988 improved the export system's overall effectiveness and efficiency. In addition, the system could be improved by establishing collective import quotas, registration of dealers, and inventories of existing stocks. Moreover, the control system was grossly underfunded. With adequate funding from the parties and NGOs, the system could achieve its mandate.[27]

After a three-year study of trade practices, the Ivory Trade Review Group (ITRG), an *ad hoc* group of specialists mostly from and funded by environmental NGOs, including Trade Records Analysis of Flora and Fauna in Commerce (TRAFFIC)[28] and the Worldwide Fund for Nature (WWF) concluded that CITES controls were not fully responsible for the decline in total ivory trade volume. Rather, smuggling, stockpiling, population reductions, and changing public attitudes were major contributing factors. Although some information from industries in the consuming countries indicated a relatively high degree of compliance with the CITES regulations, the ITRG criticized the CITES system as one which was easily evaded.

On June 1, 1989, the ITRG released the findings of a study commissioned by the African Elephant and Rhino Specialist Group of IUCN. The study was funded by Wildlife Conservation International, WWF, US Fish and Wildlife Service (FWS), and was carried out by TRAFFIC, the CITES Secretariat, and the African Wildlife Foundation. The ITRG report referred to

> further evidence of the chaotic, uncontrolled conditions of the international ivory trade today. For example, it found that poaching had become so prevalent that 'the legal [government controlled] and the illegal trades have become virtually indistinguishable' and that 'exploitation of elephants to supply ivory, as currently practiced throughout most of the continent [of Africa] is quite unsustainable'.[29]

The CITES Secretariat admitted that some trade had occurred outside the quota system, but pointed out that the system was not designed to eliminate poaching within national boundaries. Anti-poaching efforts are the responsibility of the states' enforcement agencies, for which CITES is no substitute. Thus, in its 1989 evaluation of the quota system the secretariat concluded that some of the most fundamental problems, such as inadequate resources for elephant protection and management programmes, reside at the national level. If legal trade in ivory was to continue, the secretariat argued, attention had to focus on the implementation of strict trade controls with uniformly high standards.[30]

At the national level in the late 1980s, both producing and consuming countries pursued more drastic measures to save the elephant. Kenya, for example, appointed anthropologist Richard Leakey as Kenya's Wildlife Services director and gave him considerable power to reform the agency and stop poaching. Leakey immediately began weeding out corrupt officials and raised park rangers' wages. He also implemented a shoot-to-kill anti-poaching campaign, providing his wardens with automatic rifles and helicopter gunships.

Tanzania followed suit, rounding up 1,800 ivory poachers and middlemen in just a few months.[31]

With the intergovernmental institution admitting that it could do little to address the elephant problem, and with range states reduced to drastic measures, the stage was set for transnational efforts, especially by environmental NGOs. Up to this point, large conservation NGOs had participated mainly to document the decline in populations, as well as to shore up national wildlife and parks agencies. Other NGOs, many of which were oriented toward animal rights and preservation, had lobbied for at least a decade for a ban on trade in ivory, but were dismissed by CITES, its member states, and, not least, the conservation NGOs. In the late 1980s, however, the population trend appeared unmistakable and provided a rallying point for more drastic action. The preservationist NGOs thus gained a voice that few could ignore, including the conservation NGOs who eventually joined the move for moratoria and, then, a complete ban. Much of this political effort began in the United States.

A major lobbying campaign by environmental NGOs resulted in the passage of the United States African Elephant Conservation Act of 1988 (AECA).[32] Designed to halt the flow of illegal ivory into the United States, the AECA required the Secretary of the Interior to conduct an investigation of all ivory-producing states to determine which states had effective elephant protection programmes. For those states lacking effective programmes, import moratoriums would be placed on their ivory. Import criteria also would be applied to intermediary states. Although the act did not call for a trade ban, it did give the US Management Authority the authority to institute selective moratoriums against deviant countries.[33] The act complemented the NGO publicity campaigns by heightening public concern for the elephant. It also helped push up ivory prices, serving to accelerate poaching, as traders anticipated the selective moratoriums. And it accelerated the move toward a total trade ban on ivory.[34]

In early May 1989, four range states, Tanzania, Kenya, Gambia, and Somalia proposed a total ban on ivory trade. The United States did likewise, and in late May Britain joined the call for a ban. Other African states followed soon after. The IUCN and the WWF warned that poaching could accelerate in anticipation of a total ban at the upcoming October 1989 CITES meeting in Lausanne, Switzerland. Thus, these organizations encouraged trading states to declare immediate, unilateral bans on ivory trade.[35]

Shortly after the ITRG released its findings, on June 1, 1989, the United States announced a unilateral moratorium on the importation of all ivory. France, West Germany and the European Community declared similar bans.

Asian consumer countries were not as quick to shut down their ivory businesses, however. Hong Kong banned only imported worked, not raw, ivory. Japan banned raw and worked ivory, but not from African countries party to CITES. Japan also continued its trade with countries such as South Africa and Zimbabwe that had effective elephant management programmes and that opposed the ban, on the grounds that a ban would curtail revenue needed for wildlife management.

In fact, Zimbabwe led a subcommittee of CITES, the African Elephant Working Group, made up of producing and consuming countries, that met in Botswana in July 1989. That group concluded that:

1 the ITRG study was seriously flawed;
2 the number of African countries favouring a ban were in the minority;
3 trade should continue;
4 the major importing countries were also opposed to a ban.

The southern African countries also proposed a common marketing system for ivory, whereby all raw ivory would be sold by auction from a single outlet and would operate in accordance with CITES provisions.[36]

At a press conference in September 1989, Zimbabwe's Minister of Natural Resources and Tourism explained that:

> Reports on the elephant 'crisis' give the impression that, until recently, vast herds of elephant roamed the continent. We are asked to believe that in the last few years this Utopia has been destroyed by illegal hunting.
>
> This scenario is not true for Zimbabwe and many other African countries. . . Elephant products such as ivory, skin and meat have earned Zimbabwe about Z$20 million in direct exports since 1980. . . This has assisted in conservation and placed a high value on elephant.
>
> The international community should be aware that successful conservation of elephant will inevitably lead to ivory production. If we were not to manage elephant, it would be naive to believe that they would stop producing ivory. Greater numbers would die naturally, more would be killed as problem animals, and illegal hunting would increase.[37]

With unilateral bans multiplying, and opposing sides building their arguments, attention shifted to the Seventh Meeting of the Conference of the Parties, held in Lausanne, Switzerland, October 1989. The meeting was covered by almost 160 journalists and sixteen television networks. More than 70 per cent of the resulting news stories dealt with the African elephant.[38]

Although many expected a worldwide ban to be implemented, others – most notably the CITES Secretariat and the southern African states – were determined to head off such a drastic move. The secretariat argued for adequate financial support for the control system, and strengthening of the range states' elephant conservation programmes.[39] The southern African states strenuously objected to a blanket ban. In their region, governments rather than private traders had been conducting the ivory trade. With stable, even growing, elephant populations, these countries – most notably, South Africa, Zimbabwe, and Botswana – explained that they would be unduly penalized for the inadequacies of other producer countries. When their amendment to allow lawful sale of ivory in their region failed by a vote of twenty in favour, seventy against, and one abstention,[40] they threatened to take reservations[41] and continue selling ivory. Botswana even hinted at withdrawing altogether from CITES. The often rancorous debate among the CITES parties, then numbering 108, eventually resulted in an uplisting of the African elephant to appendix I. The vote was seventy-six in favour, eleven against, with four abstentions.

South Africa, Zimbabwe, Zambia, Botswana, and Malawi filed for reservations and announced they would continue to sell ivory.[42] Botswana, in fact, ordered an immediate culling of some 2,500 to 3,000 elephants out of their northern herd, which numbered more than 67,000, which, the government said, was causing widespread damage to vegetation in Chobe National Park.[43] China entered a reservation concerning ivory imports, and Britain entered a reservation on behalf of Hong Kong that allowed the colony a six-month trade extension. Japan initially abstained from the vote in Lausanne, but later agreed to abide by it.[44]

The demand for ivory began to drop within one year of the ban's implementation. Poachers found it difficult to sell ivory, and dealers were not giving poachers their usual advances to acquire as much ivory as possible. Ivory became less valuable as the demand from the rest of the world fell. Consequently, prices fell or remained stable throughout most of the world.[45] Before the unilateral ivory trade bans of 1988, wholesale prices of ivory had reached $200 per kilo.[46] After the ban, prices in Zaire dropped by more than half, and in the Congo by 20 to 50 per cent.[47]

Preliminary data revealed significant changes in Central and East African countries where poaching had been heavy. Kenyan officials reported that the price of illegal ivory collapsed there. Ivory dealers apprehended on the Somali border were attempting to sell tusks at $5–$7 per kilo. Prices in Somalia were even lower, $2–$4 per kilo, and 'There were no takers'.[48]

In the UAE, a key entrepôt and processing centre, factories were shut down and ivory sales banned. The local economic impact was reported to be minimal. There was no evidence of smuggling. Moreover, UAE joined CITES in May 1990.[49]

The ban was dramatically successful in shutting down the US ivory market. Dealers reported that wholesale prices for jewellery and simple carvings, which traditionally accounted for most of the US ivory market, were discounted up to 70 per cent. One major ivory trader in New York said that price was irrelevant in the US market, because demand for ivory was non-existent and trading did not occur at any price. In Asia, China and South Korea continued to import ivory. China did find it difficult, however, to locate either internal or external markets for its ivory stockpiles. Although Chinese ivory factories were operating at a tiny fraction of their capacity, China showed no indication of withdrawing its reservation. Much of the South Korean ivory was shipped to illicit markets in Japan. After an initial surge following the uplisting, prices and demand eventually declined in Japan.[50]

As of 1990, the trade ban appeared to have had a negligible impact on the overall economies of most elephant range countries. The southern African states claimed, however, that the inability to sell ivory products had a detrimental effect on wildlife conservation, by eliminating one source of income for wildlife management programmes. US government sources report that Zimbabwe estimated losses of up to $9 million in general revenues as a result of the trade ban. Yet South African sources indicated that, although trade bans have had an effect on their elephant population management efforts, the overall influence on the nation's economy has been insignificant.[51]

The initial success of the ban and the ability of the CITES Conference of Parties to lift it will depend in part on the availability of ivory substitutes. Steinway and Yamaha, major piano manufacturers in the United States and Japan, announced they would use synthetic ivory in their keyboards. Mammoth and walrus ivory, hippo teeth, bone, and palm 'ivory' are naturally occurring substitutes that some ivory craftspeople in the United States, Japan, and elsewhere are beginning to use. Livestock horn, wood, stone, and ceramic can also be used for Japanese signature seals.[52]

Although many characterized the trade ban as a success, others were quick to point out that it was not a solution; enforcing the trade ban and encouraging substitutes is not enough to ensure the survival of the African elephant. The African Elephant Action Plan, revised in March 1990 by the African Elephant Conservation Coordinating Group (AECCG) and funded primarily by WWF, the FWS and the European Economic Community (EEC), argued that sufficient funds and technical expertise must be made available to range states to protect and manage their elephant populations. The AECCG has helped range states prepare conservation plans and coordinate proposals for donor organizations. The 1990 plan outlines five categories of conservation deemed essential to save the elephant:

1 field action, such as anti-poaching patrols;
2 maintenance of control in international trade of elephant products;
3 coordination and management plans for elephants and their habitats;
4 public awareness and education;
5 research and survey work.

The AECCG actively solicited funds for projects in these categories from intergovernmental organizations, government aid agencies, and private conservation foundations.[53]

In April 1990, France sponsored a donors' meeting in Paris that brought together seventeen former ivory-consuming nations, the EC, WWF, and other concerned organizations to discuss how to meet the critical need for elephant conservation funding. The US Congress had previously committed $2 million in overseas assistance funds to that end. WWF alone spent more than $2.7 million on elephant projects in nine African nations in 1990.[54]

BANS, REGIME CHANGE, AND NGO LEARNING

To explain and assess this case and explore its broader implications for NGO relations and the NGO role in reversing trends in environmental degradation, I develop three major points:

1 the *ban* as an environmental tactic;
2 NGOs and CITES as a biodiversity regime;
3 international conservation and NGO learning.

The ban as an environmental tactic

Bans in international commerce and security are nothing new. Piracy, slave trading, counterfeiting, and drug trafficking have all been subject to prohibition. Similarly, many attempts have been made to limit or remove weapons of mass destruction or, as in Antarctica, to ban military activity. In the natural resource arena, bans generally have been applied when a resource is severely depleted. As such, the ban is a last-resort measure to save the resource. Examples include bans on the hunting of fur seals and whales and, now, on the trade in products of the African elephant.

On the face of it, the function of a ban is straightforward: a resource is overexploited, and it needs a recovery period.[55] The timing and implementation of a ban is not as clear, however. Moreover, it may be that a ban serves broader strategic purposes. Under conditions of rapid and irreversible decline, a ban may be useful, possibly necessary, in promoting a broader conservation and environmental agenda. To examine the broader implications of the ban as an environmental tactic, I first explore its special features, progressing from the individual and organizational levels to the national and international levels. I then turn to the implications of a ban on a biodiversity regime and on NGO relations and show that a ban may be a necessary condition, but not a sufficient one, for long-term conservation. I argue that, in general, bans may be necessary to reverse trends when traditional approaches grounded in prediction-oriented scientific management are found lacking.

When limiting an undesirable activity, the differences between zero and one unit of that activity, and between ninety-nine and one hundred units are not the same. Mathematically, they are, of course, identical. But psychologically, they are not. If I need to cut in half my chocolate intake to eliminate headaches, it is difficult to know when, after a few chocolate chip cookies, a chocolate malt, and a candy bar, I have done so. When have I consumed my usual 10 oz., when 5 oz., when 5.5 oz.? If, however, I eliminate chocolate altogether, I know for certain that I have sufficiently curbed my habit. It may not be 'rational' to do so, because I receive so much satisfaction from chocolate (in fact, I am sure it stimulates creative thinking); all I really have to do is eat less than, say, 5 oz. But zero consumption is readily identifiable; it is prominent. Unlike the infinite number of other possibilities in a continuous consumption function, it is uniquely recognizable. Zero is, in short, a focal point.[56]

This focal point quality of total abstinence applies to prohibitions of a wide sort, one being a trade ban. In the ivory case, zero trade in ivory eventually became the only tolerable level recognizable by consumers and producers alike. Producer countries with effective ranch management practices argued for limited trade. But on the world market no amount could be readily identified, even though some level was certainly sustainable.[57] Thus, although a 'rational' volume of trade could have been constructed from biological and economic data, saving the species meant that only one volume was sustainable: zero. In other words, some level of consumption may have been biologically sustainable, but not politically sustainable.

The organizational and institutional features of a ban build on the individual psychological features. Restaurants often find it easier to ban smoking outright than to set up no-smoking sections from which smoke invariably drifts and which is never the right size for optimal seating. Similarly, an absolute ban on handguns is easier to enforce for British police than the partial controls American police must deal with. Likewise, low-level bureaucrats are notorious for their strict adherence to seemingly meaningless rules. When clients request exceptions, exceptions that would benefit client and agency alike, the answer is no. All or nothing is easiest to implement because, as organizational theorists have observed, operating by the book reduces bureaucratic uncertainty and the stress of making decisions on an *ad hoc* basis.[58]

Organizationally, therefore, it is easier to just say no. With a ban on ivory, those agencies charged with protecting the elephant found it easier to shut down carving and trading enterprises, and even easier to shoot poachers, than to institute an effective management scheme or reward local communities for conservation practices. Customs agents worldwide found it easier to confiscate all ivory than to separate the legal from the illegal ivory.

At the national and regional levels, poaching and trade tends to flow to the path of least resistance, that is, to those states where legal loopholes and institutional weaknesses are greatest. A ban checks this process by interfering with safe havens and transshipment centres for poachers and illegal traders. An overall ban (even, conceivably, regionwide) is, thus, mutually reinforcing. Kenya, for example, outlawed elephant hunting in 1978. Between 1981 and 1989, however, its elephant population dropped from 65,000 to 16,000.[59] Even its shoot-to-kill policy against poachers was only marginally successful. Only when neighbouring states joined in the absolute ban did Kenya begin to reduce illicit killing.

At the international level, a ban can help create the broad-based conditions – especially norms of consumer behaviour – necessary to save a species. When anarchy (no overarching government) prevails, when international institutions are weak, and when few norms are widely accepted, effecting change in international behaviour is extremely difficult.[60] At the state level, change is nearly impossible to achieve unless the effort falls on the heels of a major cataclysm and a hegemon is prepared to accept the burdens of leadership – witness GATT and Bretton Woods after World War II, and the role of the United States. Some forms of international commerce, however, may be susceptible to changes in norms, especially those norms shared by consumers. This possibility may be greatest in the arena of natural resource trade, especially trade in the products from a popular animal like the elephant. This possibility may increasingly apply to ecosystems like rain forests, coral reefs, and wetlands.

The target of an international ban is consumers. The ivory trade ban changed consumer preferences in three ways. First, as discussed, if ivory consumers, whether these are individuals or countries, such as Japan, were asked to reduce consumption, they would ask by how much, and nobody could say precisely. The ban made it precise – to zero.

Second, consumers had little incentive to reduce their consumption levels unilaterally. Pleas for reducing ivory consumption, pleas that relied on consumers' concern for nature or guilt for killing elephants, were not enough. The collective action problem rears its ugly head. One member of a collectivity (whether an individual within a society, or a state within an international trading system) cannot rationally curtail a destructive habit if the short-term improvement is trivial and the benefits are significant for those who continue the habit.[61] A comprehensive ban solves the collective action problem in the sense that the incentives to consume, let alone to cheat, are reduced, if not eliminated.

Third, the incentives to consume are fewer because the prestige value of ivory is less with a universal ban. The ivory ban was an unequivocal statement that collectively excessive ivory consumption is 'bad'. Ivory was construed as unacceptable for carvings or signature seals or piano keys because, in the aggregate, consumption of those items led to the destruction of the species. Complemented by an intense media campaign, the outright ban associated, in effect, the ivory carving on one's coffee table with the extermination of everyone's favourite zoo and circus animal, and the spoiling of everyone's dream tour to Africa. Again, it should be noted that, as at the individual level, this solution is not optimal; some level of trade is possible and could, presumably, benefit all parties without compromising the survival of the species. But, to arrest the rapid decline in populations, regulations were not enough; changes in consumer preferences were needed, at least in the short term. Also, construing ivory consumption as bad is not equivalent to the animal rights argument that all elephant killing should be viewed as evil. Rather, I am making an argument about environmental strategy in an international climate that depreciates effective regulation, especially when extreme demand overwhelms weak institutions and peoples. Under these conditions, a change in consumer preferences is a necessary condition for saving threatened species or ecosystems.

The ban, in short, effected in one stroke a global norm that few informed consumers could ignore. Moreover, it did not require consumer conversion; consumers did not have to become ardent nature lovers or altruistic global citizens. They merely had to be convinced that the presumed benefits of ivory consumption would be overshadowed by societal condemnation of that consumption.

In general, the norm-creating effect of the ban depends on the nature of the target – namely, foreign elite consumers.[62] If only foreign elites need question their consumption behaviour to save a species or ecosystem, then subsistence use or even low-level trade need not be affected. The consequence of the norm, then, in the short term, is to temporarily halt all consumption, but in the long run to withdraw a sizeable portion of the demand so as to significantly reduce – but not necessarily eliminate – the exploitation of the resource. Therefore, although such a norm suits animal welfare activists, it is not, strictly speaking, preservationist. Also, as discussed below, the ban that produces the norm can split the NGO community and, more important, it can have differential impacts on range states and their attempts to deal with resident peoples.

From an institutional perspective, the ban strengthened a core ingredient in the CITES regime: the norms that govern international behaviour. Whereas parties, by their accession to the convention, ascribe to the norms and principles of conservation and regulated trade in endangered species, these norms are not necessarily assimilated at the consumer level. The treaty may help states to coordinate interstate commerce but, in the face of tremendous demand, demand that can overwhelm weak institutions and vulnerable peoples and eventually destroy the commerce altogether, only change at the consumer level can truly protect the threatened species. With such a fundamental change the spillover effect to other species, possibly even to ecosystems, could be significant.[63]

Thus, strategies aimed at tightening legal loopholes or increasing penalties for violations are likely to be ineffective, especially if radical, not incremental, change is needed.[64] Strategies that target behaviour at the individual level may have a better chance of success when a correspondence between the rates of environmental degradation and institutional response requires more than tinkering around the edges.[65] When a substantial decrease in consumer demand for a threatened species or ecosystem is needed, an outright ban may be the only way to coordinate behaviour, implement trade regulations, and thwart the effects of overwhelming disparities in income and institutional capacity.

With respect to the NGO role, strategies that at once target consumer behaviour and international norms are unlikely to be taken up by states or their international organizations. States pursue economic growth through the promotion of production, consumption, and trade. Range states (or, at least, elites within such states) that are earning revenues from ivory exports have little incentive, especially in the short term, to eliminate production and trade. This is especially true if it appears to put that state at a competitive disadvantage. Consuming states like Japan, Hong Kong, or the United States not only satisfy domestic demand through trade and consumption but support domestic carving and retail businesses. They have little incentive to curtail consumption, especially if the problem is distant and appears not to be of their making. Finally, if both producing and consuming states have incentives to promote – or maintain – ivory trade, the international organizations they create such as CITES do as well.

Confronted with rapid population declines and the imminent collapse of production and consumption activities, international environmental NGOs step in to fill an important strategic and diplomatic niche. Strategically, they push for a ban. They do not opt for simply better management schemes, schemes that are better funded, better managed, or better enforced, whether domestically or internationally. They realize, explicitly or implicitly, that only a ban, even if temporary, will halt rapid decline. Diplomatically, as discussed below, they operate at the international level (with CITES), the national level (with relevant agencies), and the local level (with communities in range states and with traders and retailers in consuming states). No other international actor so operates. Range states, especially those which have adopted colonial wildlife policies that exclude resident peoples for expatriate tourism, have not addressed local needs. Distant consumer states have little knowledge of the biophysical and social conditions surrounding the decline of the resource. International organizations, as discussed below, may have the knowledge but they cannot penetrate member states' boundaries. Thus, NGOs become key actors under such biophysical and political conditions because they can operate transnationally linking these conditions across local, national, and international levels.

To this point, the analysis of NGOs in the politics of international conservation has been largely static and has treated NGOs as monolithic. This is appropriate with respect to the short-term question of reversing rapid population decline. But long-term solutions will require adjustments on both the consuming and producing sides of resource use and more nuanced tailoring of policies to accommodate local conditions. To this end, NGOs and the ban have had differential effects in Africa. I return to this dynamic after considering the historical and contemporary transnational roles NGOs have played in the endangered species trade regime.

NGOs and CITES as a biodiversity regime

CITES is often cited as an exemplary international regime. One hundred plus countries of disparate interests manage to agree to regulate their trade in wildlife. National behaviour – including management practices and customs procedures – is coordinated through the functions of the secretariat, the Standing Committee, and the biennial meetings. But, whereas international regimes are typically conceived as a product of interstate or intergovernmental relations, the ivory trade case suggests that, in practice, this bio-diversity regime, its formation, its norm creation, and, especially, its involvement of non-state actors, is far more complex. And where prescriptions for improving the institution abound, their state-centric, legalistic assumptions tend to miss much of the politics of the regime, especially the transnational, non-state politics and, consequently, tend to skew the prescriptions toward tightening loopholes and strengthening enforcement.[66] The perspective developed in this case study, with its focus on NGOs and the role of a ban, suggests that to properly understand and prescribe international remedies for species loss and environmental degradation, it is necessary to conceptualize a regime in terms of multiple actors and multiple activities; traditional diplomacy among national governments, as argued in chapter 2, is only part of the picture.

In this section, then, I first briefly document the history and financing of CITES to show that, unlike the trade regime of GATT or the monetary regime of Bretton Woods, the endangered species trade regime is very much a product of non-governmental forces. I then argue that, to become a truly comprehensive *biodiversity* regime, trade regulation may not be enough; prohibition, the banning of trade, and the limiting of local exploitation may necessarily become common, if only temporary, measures.

CITES is a trade regulation regime. Its objective is not to halt cross-border exchange of plant and animal products, nor to protect habitats or ecosystems. Rather, CITES is organized to manage trade, especially to ameliorate the negative side-effects of the trade in endangered species while preserving states' rights to engage in that trade. If national policies lead to the extinction of a species, that is not the business of CITES.

Yet, if, to save species, bans are increasingly employed to halt trade temporarily and blunt local exploitation, CITES may well be changing from a limited trade regulation regime toward a 'global prohibition regime'.[67] Transforming a trade regulation regime into a prohibition regime has implications not only for endangered species under CITES but, quite possibly, for other international resource questions, including trade in tropical timber and even hazardous waste.

Viewed as a prohibition regime, CITES shares at least two features common to prohibition regimes targeted against such activities as piracy, slavery, and drug trafficking. One feature is a common evolutionary pattern where the targeted activity is first widely considered legitimate and is sanctioned by states, and then prohibition gradually becomes institutionalized.[68] Regarding ivory, for centuries both native groups and colonial powers promoted and controlled the hunting of elephants and the trade in ivory. In southern Africa, for example, in the seventeenth century the Dutch promoted elephant hunting and, then, when populations were depleted, attempted to regulate it.[69] The legitimate, state-sanctioned hunting of elephants and sale of their products continues to this day, even with a trade ban. The southern African countries, for example, still cull their herds, sell the meat, and stockpile ivory.

A prohibition regime begins to take shape when individuals and groups realize that transboundary influences make local control efforts inadequate.

Moreover, the activity is increasingly seen as a problem and, eventually, as an evil. In late nineteenth century Africa, excessive hunting aroused the concern of both hunters and conservationists in Africa and Europe. Their concern prompted international efforts to establish national parks and game reserves. European preservationists and game hunters also pressured the colonial governments to draft and sign what became the first international environmental agreement, the Convention for the Preservation of Animals, Birds, and Fish in Africa, signed by Britain, France, Germany, Italy, Portugal, and the Belgian Congo.[70] Although never implemented, this and subsequent conventions in the early twentieth century, initiated and promoted by nature and hunting organizations, became the seeds of international cooperation for wildlife management in the post World War II period.

With respect to ivory, the major institutional innovation was CITES and the CITES Secretariat, formed in 1975. Much of the debate through the 1950s and 1960s leading to its formation centred on the organization and funding of the IUCN and WWF and the role of the United Nations.[71]

The origins of CITES can be traced to the interwar period when two major international NGOs – the International Committee for Bird Protection and the International Office for the Protection of Nature (IOPN) – were established. A major focus of IOPN activity by British and American naturalists was British East Africa. Concern over dwindling populations of many species led to the signing in London in 1933 of the Convention for the Protection of the Fauna and Flora of Africa. The convention's aims were largely limited to creating national parks in colonial territories.

After the war, a number of prominent naturalists, including P. G. van Tienhoven, President of IOPN in the interwar period, Charles Bernard, President of the Swiss League for the Protection of Nature, Max Nicholson, Director-General of the Nature Conservancy in Britain, and the British biologist Julian Huxley debated whether to strengthen or replace IOPN. Although some conservationists sought to build on the IOPN's structure, in the climate of the post-war period many urged institutional innovation. At a meeting in Basle, in 1946 the question of whether the organization should be inter- or non-governmental was left unresolved, with the conclusion that 'it is desirable that there should be an active international organisation, widely international and representative in character, adequately financed and with adequate terms of reference.'[72]

Huxley, the first Director-General of the United Nations Educational, Scientific, and Cultural Organization (UNESCO), pushed for a nature protection organization. Delegates to a second international conference, this time in Brunnen in 1947, were wary of the newly formed UNESCO, which had among its tasks the job of setting up non-governmental or semi-governmental organizations. Heading off an increased role for UNESCO, the delegates instead created a provisional International Union for the Protection of Nature (IUPN). UNESCO was concerned about the degree of governmental participation in the formation of the union, so in a 1948 report called for the creation 'of an international non-governmental organisation for the preservation of nature rather than an intergovernmental organisation proposed by the Brunnen Conference.'[73] In preparation for a third conference, the French government stepped in with its own plans to upstage UNESCO. A compromise was eventually reached in which France and UNESCO jointly invited governments to send representatives to the conference, while the provisional IUPN invited private bodies.

Voting powers of the proposed union became the main issue at the resulting conference in Fontainebleau. Governments feared allowing private groups in a country to outvote their own government. The question was resolved by giving government members two votes in the union and private groups combined one vote in general assemblies. Shortly after Fontainebleau, UNESCO signed a contract with IUPN, giving it financial support. IUPN had four categories of membership: governments, agencies of governments, international inter- and non-governmental organizations, and national non-governmental organizations. Among its objectives was the drawing up of a worldwide convention for the protection of nature.

From the start, the International Union for the Conservation of Nature (IUCN; so renamed in 1956) suffered from financial problems and an inability to respond to urgent conservation needs. Out of concern for IUCN's inadequacies and especially for increasing threats to East African wildlife, Max Nicholson and others created the World Wildlife Fund (WWF) in 1961. With support from wealthy individuals and corporations and the endorsement of members of the British royal family, the launch of WWF was marked by a six-page 'shock issue' of the *Daily Mirror* with a picture of the doomed black rhino and the now famous panda logo.[74] Heading off other efforts to save African wildlife at a time of rapid decolonization, Nicholson argued that 'Europe had to be the home for the new organisation because both the scientists and the possibilities for raising money were there.'[75] Nevertheless, WWF did attempt to gain the support of Africa's new leaders. In doing so, it often appealed to the financial benefits of conservation, especially from hunting and tourism. Over the years the promotion of hunting hindered WWF's fundraising, but the promotion of tourism in national parks became a mainstay.[76]

WWF was extremely successful in raising funds. By 1967, it had supported a total of 183 conservation projects (sixty-five in Africa) totalling $2.2 million. These projects included scientific research by IUCN, as well as the establishment and improvement of national parks.[77]

In 1963, IUCN met in Nairobi to discuss international wildlife trade of threatened species. The meeting resulted in a call for an international convention to establish worldwide controls over trade in endangered wildlife and wildlife products. At its 1969 meeting in Delhi, the IUCN listed the species it believed should be controlled by the comprehensive convention called for in 1963. Included in this list was the African elephant. The IUCN prepared several drafts of such a convention. After several agencies within

IUCN examined the first draft and a second was reviewed at its 1966 Assembly, a third version was circulated to governments in 1967. Thirty-nine governments and eighteen international organizations sent back comments. During the circulation of another draft in 1971, the United States said it would be willing to convene an intergovernmental conference. Several prominent American conservationists, including Secretary of the Interior Stewart Udall and the Council on Environmental Quality chairman, Russell Train, who had strong IUCN ties, supported the effort, arguing that a trade convention would strengthen the 1969 US Endangered Species Act. The resulting convention was held in Washington, DC, with a large US delegation headed by Train and officials of many conservation NGOs. On March 3, 1973, the Convention on International Trade in Endangered Species of Wild Flora and Fauna (CITES) was signed by twenty-one of the eighty participating nations. CITES entered into force after ten states ratified it on July 1, 1975.[78]

This brief history of CITES reveals how norms and procedures were instituted to form an international trade control regime. Moreover, it reveals the prominent role of NGOs in the formation and, as will be seen shortly, the maintenance of the regime.

Once in place, CITES, like other regimes, can create significant social pressures to acknowledge and enforce the regime's norms. At the same time, as with other regulatory regimes,

> regime proponents must contend with the challenges of deviant states that refuse to conform to its mandate, weak states that formally accede to its mandate but are unable or unwilling to crack down on violators within their territory, and dissident individuals and criminal organizations that elude enforcement efforts and continue to engage in the proscribed activity.[79]

Proponents of elephant conservation contended with deviant entrepôt and processing states such as the UAE and Singapore, African range states that were unable or unwilling to police violators, and sophisticated poaching and illegal transshipment networks. Under these conditions the ivory quota system established by the secretariat in 1986 proved ineffective. Although some argued that the quota system could be tightened, its fatal flaw – voluntary 'quotas' – was built in from the start, and for good reason. This flaw is particularly revealing of the nature of this regime and the role NGOs play.

The CITES regime – or, more specifically, the secretariat – walks a thin line as an organization. On the one hand, it has the charge of protecting the world's traded endangered species through a system of trade permits. On the other, it must accommodate one hundred-plus countries, each with special interests and capabilities regarding wildlife management and trade.

Accommodation is, in part, achieved by the several exceptions to compliance built into the convention, including 'reservations' to uplistings. Accommodation is also achieved by drawing a clear line between CITES's authority and national authority. As shown in the case history given, the secretariat frequently reminds its critics that its mandate is *trade*, not the internal management policies of the respective governments.

Conservationists and legal scholars readily denounce these limitations as serious weaknesses in the convention. The exceptions do, indeed, give an 'out' to a party. At the same time, however, they help keep each party *in* the regime. These accommodations can be understood as necessary provisions to maintain a fragile coalition. The strength of the convention and the strength of the secretariat overseeing the convention is very much in the number of parties. As the Secretariat is quick to point out, an effective *global* regulatory regime requires global participation. It does little good if half the world's countries adhere rigidly to the convention's provisions while the other half trades freely in threatened species. Thus, because the temptation to free ride is considerable, one way to prevent massive defection and, hence, the collapse of the entire regime, is to allow temporary exceptions. Of course, the more exceptions there are, the weaker the regime will be.[80]

CITES's delicate balancing act, therefore, is to attract and keep the maximum number of parties while minimizing exceptions to the rules. This is an institutional imperative. It is a tension CITES must deal with in an anarchic world, a world without overarching authority and without effective enforcement mechanisms for *any* kind of international transgression. The CITES arrangement is not necessarily best for protecting endangered wildlife, however. The secretariat may accommodate a large number of governments and work hard at maintaining good relations with all parties (much like any foreign ministry), but endangered species do not necessarily benefit.[81]

The need to stringently respect national boundaries, to avoid criticism, and to seek accommodative, not confrontational, approaches to its trade policies results in a cautious approach. On some matters this may be the wise course. But in situations of urgency, when species are on the verge of extinction, the operational imperative to exercise caution and allow flexibility in the rules can thwart the ultimate goal of the regime – namely, to protect endangered species.

The CITES Secretariat knows this tension very well, of course. As individuals, members of the secretariat are as committed as any to saving endangered species. Caught in this bind they look for an out. That out, it appears, is extensive use of NGOs, as the secretariat itself explained in an annual report:

> The Secretariat is in permanent contact with a very large number of non-governmental conservation organizations at regional, national and international levels and is fully aware of the indispensable role played by these organizations in achieving the objectives of CITES. . .
>
> One organization that must be mentioned, however, is IUCN – the World Conservation Union. This body has been of enormous help to the Secretariat, especially in the scientific and legal fields. . .To improve the co-ordination of activities with this organization a tri-lateral meeting, including WWF-International, is organized every three months.

Finally, the contribution made by the TRAFFIC network in South America, Japan, Europe, Oceania and the United States of America must also be mentioned. The information, studies and, in certain cases, assistance in the field provided by the TRAFFIC offices justify the special status of these NGOs *vis-à-vis* the Secretariat.[82]

The secretariat's close relationship with NGOs is the result, in part, of the parties' failure to meet their financial obligations or to budget sufficient funds to cover the secretariat's mandate. For example, according to the secretariat, an ivory trade study called for by the parties made no provision for financing, thus, forcing the secretariat to seek external support.[83] As a result of situations like this, the secretariat commits considerable effort to raising funds. In this respect, the secretariat itself acts much like an NGO.

Kosloff and Trexler trace the unusual degree of NGO participation to provisions in CITES. The convention not only explicitly permits NGOs to participate as non-voting observers at the biennial meetings, but they also receive, as registered observers, all documentation pertaining to the upcoming meetings. NGOs attend plenary sessions and most committee meetings. And, as it turns out, NGOs, both conservation and trade-oriented groups, contribute considerable time and financial resources to CITES for enforcement and implementation. They have developed publicity materials for CITES, printed export permits for Bolivia and Paraguay, done population studies on a number of species, and conducted training seminars for officials from the management authorities of less-developed countries. Many conduct their own investigations of illegal trading. On one occasion, NGOs even paid the expenses of more than thirty delegates to the fifth biennial meeting.[84] After that meeting, the US Government Accounting Office (GAO) was asked to investigate charges that NGO payments for travel and expenses of delegates were intended to influence their voting. The GAO found that they did not.[85] NGOs have also paid for certain secretariat activities and studies and contributed to implementation seminars for enforcement officials. Kosloff and Trexler conclude that

> most fundamentally. . . NGO oversight of Parties' implementing actions under CITES has been a key variable in achieving whatever success CITES has achieved. In the absence of NGO participation, CITES would very likely have followed the route of many other international wildlife measures into obscurity.[86]

In the early years of CITES, NGO representatives were often members of official delegations. Now, NGOs either attend meetings or are briefed afterward by their national delegation. One reason given for the lack of official representation is the diversity of groups and their inability to agree on a single representative. At the biennial meetings, NGOs are given space to distribute literature. In these meetings, they raise a card to request to speak, and often the chair will call directly on the prominent NGOs.[87] In the 1989 meeting, for example, one resolution asked the parties to call upon UNEP, IUCN, and TRAFFIC to provide nominees to serve on a panel of experts to review applications for reservations to the ivory ban.

It thus appears that the CITES secretariat turns to international environmental NGOs such as IUCN, WWF and TRAFFIC because they hold peculiar advantages. Unlike the secretariat or the parties, NGOs are not bound by national boundaries. They are accountable not to an electorate but only to their membership and, then, only insofar as membership and donations are maintained. They do not have to be nice to anyone. They can be, and often are, in the business of monitoring, exposing, criticizing, and condemning. They need not compromise on either ecological or ethical principles, or, at least, they need do so much less than governments for which the essence of maintaining good relations is, indeed, compromise.

For all these advantages, by themselves NGOs can do little more than raise public awareness and fund a park here and there. To address major problems in a systemic way – ecosystemic and international systemic – NGOs must engage the relevant systems. At the international level, this means striking deals with governments and intergovernmental bodies. In the case of CITES, this means trading on their special advantages to meet the needs of the secretariat while exploiting the secretariat's advantages. The bargain looks like this.

To resolve the tension between maintaining the coalition and enforcing rules, the CITES Secretariat plays a two-way game. With the parties it seeks good relations. With the NGOs it seeks frank assessments and investigations of the parties and, when necessary, public condemnation of deviant state practices. It is a variation on the good-cop-bad-cop routine, in which the secretariat can have it both ways – good relations as well as strident condemnation. Of course, there is a price. The NGOs demand, and get, access. Unlike comparable bodies, for example, GATT or IWC,[88] NGOs are prominent players in the biennial meetings, in the preparatory sessions for the meetings, and in the studies and reviews of the biological and trade status of endangered species. They do not just sit in the lobbies holding signs. They have a seat at the table and are called upon and looked to for information and support on a given position.

The inter-organizational exchange, then, is between the institutional imperative of the secretariat and the political needs of the NGOs, between resolving the coalition-maintenance/condemnation tension and ensuring a single-minded species protection focus, between international cooperation and international enforcement. Viewed this way, the endangered species trade regime is not the product of interstate decisions served by a secretariat. Rather, this regime is the resultant of interlocking political acts, including those of the parties, the secretariat itself, and, not least, the NGOs.

NGOs do more than provide the secretariat with information and single out deviants. In the broader context of building a stringent trade regime where bans are commonplace, they perform the role of what Nadelmann calls 'transnational moral entrepreneurs.'[89] The pervasive feature of international relations is the lack of a central authority and, consequently, weak enforcement mechanisms. To the extent that cooperation exists, norms of international behaviour are of paramount importance. Global norms are not only those encoded in conventions and treaties, but those existing in the implicit rules and patterns that govern behaviour of states and non-state actors, including producers and consumers. The evolution of prohibition regimes depends not only on traditional security and economic interests, but on moral interests as well. As definers and purveyors of moral interests, international NGOs are key actors.

> These groups mobilize popular opinion and political support both within their host country and abroad; they stimulate and assist in the creation of like-minded organizations in other countries; and they play a significant role in elevating their objective beyond its identification with the national interests of their government. Indeed, their efforts are often directed toward persuading foreign audiences, especially foreign elites, that a particular prohibition regime reflects a widely shared or even universal moral sense, rather than the peculiar moral code of one society. Although the activities that they condemn do not always transcend national borders, those which do go beyond borders provide the proselytizers with the transnational hook typically required to provoke and justify international intervention in the internal affairs of other states.[90]

Benjamin Franklin and Lord Nelson were early opponents of state-sanctioned piracy; the Quakers were among the first to oppose slavery; and American missionaries led the campaign against opium. In the trade in endangered species, conservation and animal rights groups and prominent individuals such as Richard Leakey and the Prince of Wales led the effort to save the elephant. On an ongoing basis, TRAFFIC was almost the sole monitoring agent of ivory trade. The African Elephant and Rhino Specialist Group and the Ivory Trade Review Group were mostly WWF-funded and operated. The non-profit Environmental Investigation Agency (EIA) in London uncovered and exposed some of the shadiest illegal dealings as well as the full extent of the trade and poaching. In fact, some felt that the EIA was primarily responsible for arousing worldwide attention and turning the tide to protect the elephant against weak trade regulation.[91]

The experience of the elephant and, for that matter, the rhino and the estimated five to one hundred species disappearing daily suggests that, for species threatened with extinction by hunting or habitat loss or pollution, a trade regulation regime as currently practised is not enough. CITES will increasingly employ bans as an essential trade policy tool. Bans, even if temporary, will be necessary to establish norms of international conduct consonant with the urgency of species loss and environmental degradation extant today.

This is not to say that all exploitation of a threatened species or ecosystem must be condemned and brought to a permanent halt. Rather, the aim is to halt *irreversible* processes; moral condemnation must be directed at *excess* – whether excess consumer demand, trade, or harvesting – that exceeds ecological capacities. It is the extermination of species or ecosystems that must be targeted, not the sustainable use of those species or ecosystems. Considerations of sustainable use, as discussed in the next section, further qualifies the ban as an environmental strategy and raises questions about the production side and the impact of a ban at the local level.

All this is one way of saying that if technical fixes and global management schemes are insufficient for addressing the global environmental crisis, then social learning, as discussed in Chapter 3, is necessary. In particular, social learning requires a change in values for consumers which, in turn, requires effective agents of change. Moral entrepreneurs in the form of international environmental NGOs will be essential actors in the learning process. But they must do more than exhort. They must investigate, expose, condemn, and, at the same time, work with relevant governments and intergovernmental agencies. They must weigh the impact of their actions at both the international and local levels. Consequently, they, too, face a delicate balancing act. How they perform and the lessons they learn from cases like the ivory trade ban will be essential elements in the overall social learning process.

International conservation and NGO learning

The politics of the ivory trade ban raises fundamental questions about the nature, purposes and strategies of environmental NGOs. It also highlights a division within the NGO community, a division that has implications for wildlife management at all levels. In early 1990, an exchange – part public, part private – between two prominent conservationists illustrates the tensions. I quote at length to fully capture the substantive content as well as the passion of those whose lives are dedicated to preserving threatened species. Simon Lyster, an authority on international wildlife law, is senior conservation officer for WWF United Kingdom. Bill Clark is an Israeli biologist. Both were delegates to the 1989 CITES meeting that moved the African elephant to CITES appendix I, banning the trade in ivory.

In a January 1990 article in the British journal *BBC Wildlife*,[92] Lyster reviewed the success of the ban and the expected increased funding for conservation programmes in Africa. Noting the problem of exemptions to the ban, he explained that

> WWF would have been willing to support the exemption from Appendix I of those elephants in Zimbabwe, Botswana and South Africa provided those three countries agreed to a moratorium on trade in ivory until the next CITES conference, due in 1992. We insisted on the moratorium because a period of total prohibition on trade is essential to break the back of the vicious poaching. . .

But we must rebuild bridges, especially with the southern Africans. It is inevitable that the ivory will start to mount up from their elephant management programmes. If we want these programmes to continue – and we do, because they are the most successful in Africa – we must hold out the prospect of a limited, tightly controlled trade in the resulting ivory. If we do not, we will be doing the elephant no favours in the long term.

In response, five CITES delegates including Bill Clark wrote:

CITES has struggled with the ivory issue for more than a decade, and at every turn, the ivory dealers have exploited each loophole. The net result is more than a million elephant carcasses littering the African continent. All of the CITES Secretariat's indelible marking systems, and computerised quota-control schemes with sophisticated feed-back monitor systems, have failed miserably.

We saw this tragedy develop over the years. Time and again, starting with the Ghanan proposal in 1976, people who truly care about the future of the African elephant have proposed a complete ban on trade in ivory. And time and again, this has been opposed by WWF, with its bizarre dictum that wildlife must whenever possible be required to help pay for itself (on a sustainable basis!).

At the 1987 CITES meeting, when the terrible destruction of the African elephant was well known, and the WWF-collaborating African Elephant and Rhino Specialist Group (AERSG) estimated a continental population of 764,410 decreasing at 9.3 per cent a year, WWF continued to advise against a ban on trade in the elephants' ivory. 'The ivory quota control system is too new', WWF told us. 'Give the system more time to bring the situation under control.'

WWF persuaded sufficient CITES delegates, and so the system was continued for another two years, during the course of which at least 100,000 elephants, perhaps as many as 200,000 were butchered. Who will hold WWF accountable for giving such catastrophic advice? Can people who are seriously concerned about the future of the African elephant continue to give credence to WWF advice?

Last summer, when all the world was clamouring for an end to the ivory trade, the Ivory Trade Review Group (ITRG) – largely financed by WWF – estimated there were 609,000 elephants left in Africa. WWF responded as if such a report was revealing shockingly new information. Hasty press conferences were called. Demands were made for an emergency halt to the ivory trade. An atmosphere of crisis was created.

If due cognisance had been given to the 1987 figure of 764,410 elephants, declining at an annual rate of 9.3 per cent over two years, it could be predicted that approximately 628,841 would remain in 1989. That figure is, in fact, just about 3 per cent greater than the ITRG estimate.

So why the crisis atmosphere to an estimate predicted two years earlier? And why the emergency shift of WWF public orientation on the ivory issue with the publication of the ITRG report if its contents were indeed predicted with such accuracy?

We think the answer may lie inside the cash box. Last summer and autumn, campaigning for elephant protection was very fashionable and lucrative. WWF wanted to appear on the proper side of the fence. But all the while, WWF was prepared to undermine the entire ivory ban by accommodating southern African states which, despite Mr. Lyster's claim that they have 'good management programmes', are actually profoundly involved in trafficking contraband, manipulating habitat and distorting the truth.

During the October CITES meeting, it was made very clear by a number of speakers that elephant poaching was being efficiently organised by master-criminal syndicates with access to sophisticated techniques. Field poaching units were being equipped with vastly superior fire power, covering a wide range of automatic weapons. This, together with excellent transportation (sometimes in convoys) and sophisticated radio communication, put the national parks protection staff at a severe disadvantage, as the terrible elephant mortality figures adequately confirm.

In a private letter to Clark, Lyster replied:

What a world we live in. Since CITES, we have been subjected to some vitriolic criticism from the southern Africans for being responsible for the Appendix I listing, for hastening the decline of the elephant in southern Africa, and for abandoning our conservation principles. It seems WWF can never win!

But your letter did sadden me a bit. It is such a pity when conservationists attack each other in print; the public get very confused. Also, I think you rather misinterpreted the point I was trying to get across – or perhaps I did not explain it very clearly.

WWF, as much as anyone, wants to see as many elephants as possible in Africa. We also want a complete and total moratorium on all trade in ivory. We would not support any lifting of that moratorium unless the controls are adequate and meet the criteria agreed in Lausanne. In the article I was merely trying to reflect the olive branch that those criteria offer. I did not hear you, or anyone else, speak against the criteria in Lausanne so do not quite understand the vehement criticism of me now.

When I was talking about 'good management programmes' in Zimbabwe etc., I was referring only to their ability to manage elephants in their countries, not about their involvement in ivory illegally traded elsewhere. Of course the latter is highly relevant to whether or not they should be allowed to trade in future (as the criteria agreed in Lausanne make very clear), but surely you would agree that Zimbabwe has managed its elephants well in the last 20 years and much better than almost anywhere else in Africa? Would you not also agree that the wildlife utilisation projects on marginal lands outside the parks and reserves in

Zimbabwe are better for wildlife than the alternative land uses of cattle or crops?

As I said in the article, the massive commercial ivory trade of recent years should remain a thing of the past. But if we tell the southern Africans that they should never again be allowed to trade in any ivory, even in very limited quantities, would this not be a serious disincentive to conserve elephants on lands outside parks and reserves? If we said we were in favour of a *permanent total* ban, I wonder if this might result in less rather than more elephants.

Finally, in private correspondence, Bill Clark elaborated on his views of ivory trade, conservation, and the role of the lead NGOs.

Our letter was a response to WWF policy, particularly that of taking the initiative in rebuilding relationships with southern African countries who presently defy CITES, hold reservation on the African elephant, and put at risk other elephant populations elsewhere in Africa. Don't complain about their 'vitriolic criticism.' Consider it an honour!

Don't be 'saddened' by our statement. It was meant as a cathartic. WWF needs a good scolding, and a reorientation. It has been drifting in the wrong direction deluding itself, the conservation community and the general public.

I, for one, do not necessarily want to see 'as many elephants as possible' in Africa. It is not a matter of sheer numbers. Rather it is a matter of the lasting integrity of ecological dynamics. I'd prefer a varying number of protected elephants fluctuating within the natural constraints of their unmanipulated habitats – what Caughley refers to as a stable limit cycle. Or, more frankly, we humans should keep our bloody hands off both the elephants and what's left of their habitats.

I do not hold Zimbabwe's elephant management schemes in such high esteem as you appear to. Zimbabwe treats its elephants like battery hens. Zimbabwe is only concerned with numbers and profits.

Consider for example all those artificial watering sites in Hwange National Park. Do they not serve to stimulate the elephant population artificially by eliminating the major natural limiting factor (drought) which otherwise would function to inhibit population growth? Do they not serve to attract other elephants, and from considerable distance (even neighboring countries) – I recall your lawyers consider the phenomenon an 'attractive nuisance.'

Zimbabwe provides artificial watering sites in an arid ecosystem throughout the six-month season of no rains, and then they complain of too many elephants? And their response to this artificial stimulation of population is an equally artificial reduction – culling.

All the meat from those horrid cullings is sold at profit. They sell it on a commercial market and make a very good profit. They sell it at special rates to crocodile farmers (the flesh of one Appendix I animal to feed yet another Appendix I animal, so it can in turn be slaughtered – and for what? An exotic leather market which is as disgusting as the ivory market? Dare anyone call this conservation?)

Some elephant meat is *sold* to rural communities where malnutrition is a problem. I have not read anywhere that the government of Zimbabwe actually gives to the needy.

And, of course, they sell all those elephant calves which their expert marksmen avoid shooting during culling operations. How many of those tragic infants, after having their entire family slaughtered and butchered before their eyes, are then captured and sold to bondage in menageries, circuses and other repulsive businesses? The 'good' in your reference to 'good management programmes' certainly cannot be a moral judgement which distinguishes good from evil.

Zimbabwe could continue with its habitat manipulations and sickening culls and still keep the entire process a very profitable domestic operation. But they refuse. Zimbabwe's greed insists upon pursuing international trade even when it is terribly clear that this international trade has been responsible for the catastrophe which has shattered the African elephant across the greatest portion of its range.

We have already seen consequences of the UK reservation. After a period of significant decline, there has been an upturn in poaching around Africa. The Kenyans have caught a few of these criminals and, under interrogation, they have admitted that it is possible for them to launder their ivory into Hong Kong, and from there it can be sold.

If poachers can introduce ivory into British-controlled Hong Kong, why couldn't they introduce ivory into any market set up to accommodate southern African ivory. Has not the entire sequence of CITES ivory controls always been avoided, subverted and ignored by the dealers? Is there any hint that southern Africa, already profiting from enormous trade in contraband ivory, would design a 'fool-proof' marketing system which would keep illicit ivory out? Southern Africa has long been an entrepôt for contraband ivory. Their elephant management schemes are utterly repulsive. The ivory moratorium should not be 'complete and total' as you say. It should be 'complete and permanent.'

I am unfamiliar with any criteria established in Lausanne concerning what might be considered acceptable controls for ivory. Might you kindly send me a copy?

Further, I do not agree that 'the wildlife utilisation projects on marginal lands outside the parks and reserves in Zimbabwe are better for wildlife than the alternative land uses of cattle or crops.' Most of that land is unsuited for cattle (tsetse fly), and for crops (too arid). I suspect most of it would lie as *de facto* nature reserve simply because tsetse fly eradication programmes, and agricultural irrigation programmes, are simply too expensive.

Also, I am very disturbed by WWF's financing various trophy hunting schemes on these lands (WWF Project 3749, Jan. 1988–Dec. 1992,

Expenditure as of last July U.S. $387,356). That must be about a quarter million sterling!

In his article 'Wildlife utilization and sustainable land use in Africa' (WWF Reports, June/July 1989) Russell Taylor is quite critical of 'pressure against consumptive use of wildlife.' And then he goes on to describe the glories of the WWF-sponsored multispecies project which involves all sorts of safari and trophy hunting, cropping wildlife for meat, game 'ranching', 'intensive management of confined populations of a few species' (game farming), and 'running cattle or other domestic livestock with wildlife.' This is the conservation of nature?

The Guruve District project 'provides lucrative recreational big game hunting' he writes. In the Chipinge District, adjacent to a national park, 'the community has voluntarily moved off an island on the Save River to permit its use as a safari hunting area.' In Gokwe District 'people at Nenyunga are ready to fence off a large tract of wild country for wildlife utilization, primarily safari hunting.' Should WWF support fencing wild areas for the benefit of safari hunters?

You have only to thank the voices of moderation among my colleagues who co-signed the letter and who also have restrained me from a much more vigorous complaint. I believe WWF's elephant and Zimbabwe policies are obscene from both an ethical and biological point-of-view.

You have asked that we continue the debate. And I think this is useful and healthy for the elephants. Thus, this letter. But I must confess, I am profoundly disappointed by WWF, and I shall not hide this fact.

I hasten to remind you that my involvement with elephant protection goes back more than a decade. At the 1979 CITES meeting in Costa Rica, I circulated documents calling for a transfer of the species to Appendix I and a halt to the ivory trade. I was ridiculed by, among others, WWF. I persisted through subsequent CITES meetings, and elsewhere, for ten difficult years. And through that period, it became increasingly obvious to me that the most serious obstacle to my efforts was the triumvirate of WWF, IUCN and TRAFFIC.

This trio of lettered clones has blocked my participation on AERSG. It has blocked my participation on AECCG (direct orders from Buff Bohlen, I was told). And because these Three Ugly Sisters are key participants in the organisation of the Paris donors conference, it appears that my participation there may also be precluded. Not that I don't have something to say – I'd like very much to make a few contributions regarding criteria for expenditure of funds. So it appears that I shall have to make my voice heard in other ways.

WWF was among the last to accept a halt to the ivory trade. WWF has resisted this for years. And now WWF proclaims itself the victor in stopping the ivory trade at CITES. The UK is not the only hypocrite in the campaign to protect elephants from a stupid and avaricious trade.

If there was organizational learning on the part of WWF and other conservation groups, it may have been as simple as Lyster concludes: you cannot please everyone. But internal studies conducted by WWF through the 1980s suggest that some profound soul-searching took place.[93] Those studies are primarily aimed at WWF policies in specific countries. The preceding discussion regarding the ban and a prohibition regime suggests that many actors are re-examining the international dimension of conservation as well. Together, the ban and its aftermath suggest that broader social learning has been spurred by these events and, in particular, by the environmental NGOs. Three points about social learning prompted by NGOs thus emerge from this case and the Lyster-Clark debate: the political expediency of a single species focus; the ambiguity of the biological and social impacts of a trade ban; and the domestic–international linkages in conservation.

Governments and NGOs alike find a single-species, publicly prominent strategy like a ban politically expedient. To illustrate, Kenya's heightened attention to the elephant in the late 1980s occurred, not coincidentally, at a time of declining tourist revenues and growing political instability. The elephant provided a useful focal point for both domestic and foreign political manoeuvring. The NGOs, likewise, quickly jumped on the bandwagon with major media campaigns and fund-raising efforts.[94]

The ban creates a dilemma for environmental NGOs, however. A single-species focus is untenable as an overall conservation strategy because it does nothing to address underlying causes of population decline. It is, nevertheless, useful, possibly necessary, to attract attention, to goad recalcitrant agencies, to lobby legislatures – not to mention to raise funds. And charismatic species are especially handy for these purposes. WWF's first save-an-animal campaign was for the rhino. When that got minimal response, they emphasized the panda even though it was not threatened. WWF founders originally chose the rhino because they did not want people to think of WWF as just a 'save a cute animal' organization. What they apparently quickly learned was that, although the principle may have been ecologically and ethically correct, it was not politically expedient. The panda – and, subsequently, the Bengal tiger, the gorilla, the elephant and many others – were necessary to rally attention, call for action, and, not least, support the organization.

In sum, even if an NGO's aim is to address the broader ecological and social realities of species loss, an important strategy is to find charismatic species (or equivalents like rain forests). That is, for organizational and political reasons – fund-raising, media attention, access to politicians and government officials – high-profile, emotion-laden, single-focus campaigns are necessary, even if the goal is more integrative or more 'rational'. Owing to the ambiguities of international policies and local conditions, however, the strategy carries risks.

Thus the second point regarding social learning prompted by international environmental NGOs in this case relates to the biological and social impacts of the ban. As noted, from a conservation management perspective, a single species or even a habitat focus is inadequate to protect endangered species. Susan Lieberman of the United States Fish and Wildlife Service points out that the elephant question overwhelmed everything else at the 1989 CITES meeting. The elephant may be a keystone species, but all the attention it received was at great cost, as many other traded and threatened species were neglected for lack of time and money. When loss of habitat is the greatest threat to species and the most intractable of conservation problems, a trade ban accomplishes little except to buy time. The ban was not, Lieberman stresses, an ecosystem approach to wildlife protection.[95]

The ban was also not a *social* approach to wildlife protection. Problems of excluded residential peoples or wildlife destruction of crops are not addressed by the ban and, arguably, are exacerbated by the ban. As a result of the ban, Kenya, for example, has become increasingly dependent on foreign assistance and tourism to maintain its parks and to protect wildlife. Officials and conservationists are beginning to question whether such dependence can be sustained financially. And, socially, these policies of exclusion engender local resentment among residential peoples which easily translates into increased poaching and encroachment on protected lands. Zimbabwe, by contrast, has had its revenue from ivory trade cut off, thus jeopardizing its experiments with local control of wildlife. Without those revenues, resident peoples are more inclined to drive out wildlife to protect crops or to convert wildlands to agricultural uses.[96]

In light of these social impacts, the ambiguous effects of the ban raise difficult strategic questions and these questions, in turn, contribute to social learning beyond that experienced on the consumer side of resource use. NGOs – whether environmental, conservation, development, or animal rights – are increasingly coming to see that only systemic approaches – biologically systemic and socially systemic – have long-term viability. A consensus within the international conservation community is emerging that recognizes that programmes to reduce species loss must be tailored to conditions of the ecosystem, the host country and the local community. The moral objections raised by Bill Clark notwithstanding, sustainable use – as opposed to total preservation – may require culling operations and ranch management techniques. Sustainable use may require that people and communities at the local level become the primary decision-makers about their own resources, not foreign officials and international conservationists. Although some observers may conclude that international NGOs should withdraw from local projects, this case suggests a different lesson.

Thus the third point regarding social learning prompted by international environmental NGOs in this case relates to the need for NGOs to operate increasingly at the nexus of the domestic and the international levels. Their primary function is, indeed, to link the two levels.[97]

It has become commonplace to hear that global environmental problems are, fundamentally, local problems. In the ivory trade case, WWF argued all along that it was the institutional inadequacies of many range states that threatened elephants. If these countries' natural resource and parks agencies had better training and equipment and manpower, they felt, the poaching could be halted. Moreover, they argued, if the proceeds of wildlife preservation could be realized domestically rather than by middlemen, people would have an incentive to conserve the resource, not over-exploit it. These are familiar themes in the conservation/development literatures and came out in the Lyster–Clark debate.

What this case suggests, however, is that domestic development is not enough, that a key to ensuring environmentally sound practices is to make linkages to the outside world. If the primary linkages are to traders and entrepôts with huge sums of money, domestic practices will emphasize maximum short-term exploitation. Or, if the primary linkages are to fickle tourist and foreign aid dollars, then the resource will be easily sacrificed when these revenues dry up. If, however, the primary linkages are to actors who simultaneously moderate consumer demand (through public education or, under conditions of urgency, through boycotts and bans), regulate trade (through lobbying officials and monitoring the trade themselves), and providing support for locally tailored institution building, then resource exploitation is more likely to take a long-term, sustainable path.[98] In the ivory case, environmental NGOs, in effect, supplanted commercial links with conservation links. As discussed, the sustainability of these is still questionable. But the previously dominant linkages are unquestionably unsustainable.

In short, action at either the international level (especially, for example, instituting bans) or at the domestic level is insufficient when both international and domestic forces drive over-exploitation. Effective NGO strategies must replace linkages that promote unsustainable exploitation with those that promote sustainable ones. Under some conditions this will mean breaking or significantly rearranging ties to larger processes such as international trade or even international development assistance. The relevant conditions driving unsustainability as derived from this case appear to be global disparities in income (e.g., Japan compared with Zaire), advances in production technologies (automatic weapons for hunting, transport mechanisms that circumvent legal trading practices), fluid and often illicit trade patterns, and ineffectual or corrupt governments among producers and consumers. These conditions overwhelm local institutional structures, whether at the village or national level.

Changing these linkages requires a peculiar kind of actor, one that can interrupt the destructive processes at their source. In the trade in endangered species, governments will, of course, be best able (although not necessarily inclined) to conduct investigative and undercover work and to close illegal trade routes. But no one can rectify the income disparities that generate the demand and tempt the poachers. If demand is to be changed, it must be by changing consumer preferences. Local governments and non-governmental groups in producer countries are generally powerless in the face of the vast wealth of industrialized or newly industrializing states. When states act primarily to promote consumption, others must act to change consumer preferences.[99] Thus, political space opens for those who have expertise, who appeal to higher values,[100] who command the public's attention, and who enjoy legitimacy when governments and traders do not. International environmental NGOs can step into this space (and, in part, create it) by linking domestic needs to international imperatives.

To conclude, the ban is more than a last-resort, save-a-species tactic. Under conditions of urgency, when environmental problems are characterized by irreversibility and non-substitutability,[101] it may be an essential means of promoting global social learning.[102] It may be the only way to mark a practice as unacceptable when, from the consumer's perspective, that practice appears perfectly acceptable. The difficult task – both analytical and strategic – is to target those practices that are unsustainable and retain or promote those that are sustainable.

As for institutional reform, when CITES only regulates trade and has minimal impact, a more comprehensive global biodiversity regime must be devised to penetrate national boundaries to ensure habitat and ecosystem protection. Prohibition – even if limited in time or scope – may be the only way to do that. And those actors best able to operate transnationally are not likely to be governments or intergovernmental bodies. Rather, international environmental NGOs are likely to best fill this niche.

NOTES

The case history portion of this chapter was written with assistance from Cathy Diekmann, Elizabeth Owen, and Jonathan Putnam. Support was provided by the Program on the Analysis and Resolution of Conflicts and the Faculty Instructional Grant Program, both of Syracuse University and by the School of Natural Resources and Environment, the University of Michigan.

1 *World Wildlife Fund Letter*, 2 (1), 1989; Washington, DC: World Wildlife Fund.
2 Sarah Fitzgerald, *International Wildlife Trade: Whose Business Is It?* (Washington, DC: World Wildlife Fund, 1989) 62.
3 Philippe J. Sands and Albert P. Bedecarre, 'Convention on International Trade in Endangered Species: The Role of Public Interest Non-governmental Organizations in Ensuring the Effective Enforcement of the Ivory Trade Ban', *Boston College Environmental Affairs Law Review* 17 (Summer, 1990) 807.
4 *World Wildlife Fund Letter*.
5 Douglas H. Chadwick, 'Elephants – Out of Time, Out of Space', *National Geographic*, 179 (5), May 1991, 14.
6 *World Wildlife Fund Letter*, 4–5.
7 Chadwick, 'Elephants', 13.
8 Fitzgerald, *International Wildlife Trade*, 65.
9 *The Economist*, July 1, 1989.
10 David S. Favre, *International Trade in Endangered Species: A Guide to CITES* (Boston, MA: Martinus Nijhoff, 1989) 121.
11 Chadwick, 'Elephants', 25.
12 Ibid.
13 Ibid., 44.

14 Early in the twentieth century, there were approximately 50,000 billiard rooms in the United States and more than 300,000 tables in use. To supply these tables, 60,000 balls were needed annually, requiring more than 10,000 elephants to be killed. Tusks suitable for making billiard balls must be taken from elephants aged from thirty to one hundred years of age. In 1920 an official of the world's largest billiard ball manufacturing company offered a reward of $50,000 to anyone who could produce a substitute for ivory. *Illustrated World*, 32(6) 1920: 926.

15 Tom Milikin, 'Japan's Ivory Trade', *TRAFFIC Bulletin* 7 (3/4) 43.

16 Chadwick, 'Elephants', 44.

17 Milikin, 'Japan's Ivory Trade', 43.

18 Chadwick, 'Elephants', 44.

19 Ibid.

20 Ivory Trade Review Group, *The Ivory Trade and the Future of the African Elephant: Vol. 1.: Summary and Conclusions,* prepared for the Seventh CITES Conference of the Parties, Lausanne, October 1989, 15.

21 Ivory Trade Review Group, *Ivory Trade,* 19.

22 Edward B. Barbier, Joanne C. Burgess, Timothy M. Swanson, and David W. Pearce, *Elephants, Economics and Ivory* (London: Earthscan Publications 1990), 9.

23 Joseph Alper, 'Should Heads Keep Rolling in Africa?', *Science*, 255 (March 1992) 1207.

24 Ivory Trade Review Group, *Ivory Trade*, 23.

25 Ibid., 26.

26 CITES Secretariat, October 9–20 1989, 'Interpretation and Implementation of the Convention: Trade in Ivory from African Elephants, Strengthening of the Ivory Trade Control System', Lausanne, Switzerland, 14.

27 CITES Secretariat, 'Interpretation and Implementation', document 7.23, October 9–20, 1989, 744.

28 Through reports and other data analysis, the TRAFFIC network assesses international wildlife trade for international and national government agencies, private non-governmental organizations, and the CITES Secretariat. Its stated goal is to stop illegal wildlife trade and monitor legal wildlife trade. The TRAFFIC network, although a programme of the Worldwide Fund for Nature, is often listed as an independent NGO. Fifteen TRAFFIC offices worldwide are located in Australia, Austria, Belgium, Germany, France, India, Italy, Japan, Malawi, Malaysia, the Netherlands, Taiwan, the United Kingdom, the United States, and Uruguay. WWF funds the individual TRAFFIC offices, but TRAFFIC is also closely associated with IUCN and experts from other organizations who work to control the international wildlife trade. Each office sends its own representatives to the biennial CITES meetings. WWF Factsheet, 'Monitoring Wildlife Trade – The TRAFFIC Network', March 1989; TRAFFIC-USA newsletter 11 (2), January 1992.
    According to Michael O'Connell of WWF-US, TRAFFIC-US maintains contacts with traders and shippers and asks them directly for information. Interview, Washington, DC, 1991. TRAFFIC-Japan also works with industry but devotes much of its efforts to monitoring import and export statistics from Japan's Ministry of Finance. Confidential interview, Tokyo, 1992.

29 US Fish and Wildlife Service, *Federal Register*, 54 (110), June 9, 1989, 24758–61, 'Moratorium on Importation of Raw and Worked Ivory From all Ivory Producing and Intermediary Nations.'

30 CITES Secretariat, 1989, 'Interpretation and Implementation of the Convention', document 14–19.

31 Chadwick, 'Elephants', 26.

32 The federal Endangered Species Act of 1973 (ESA) had prohibited US importation of any 'endangered' species listed on the ESA register. Since 1978, the elephant

had been listed as a 'threatened species', permissible for importation. Such imports, however, were restricted to ivory arriving from CITES party states. The regulations further required that imported ivory be accompanied by an export permit from the country of origin, even if the ivory had entered the market from an intermediary country where it had been worked into a finished product. ESA 16 USC 1538, cited in Michael J. Glennon, 'Has International Law Failed the Elephant?', *American Journal of International Law* 84, (1990) 13.

33 Susan Lieberman, US FWS, interview, 1991.
34 Environmental NGOs were instrumental in pushing the bill through Congress. US FWS, for example, received 75,000 postcards. The NGOs were also effective in getting funds directed to elephant conservation. When Congress failed to appropriate the funds authorized under the Act, the NGOs got $4 million of USAID's $18 million biodiversity programme earmarked for elephant conservation. Kenneth Stansell, US FWS, 1991.
35 US Fish and Wildlife Service, *Federal Register*, 54 (110) June 9, 1989, 24760.
36 Minister of Natural Resources and Tourism, Zimbabwe, 'Current Developments Regarding the Proposed Ivory Trade Ban', press release, September 1989.
37 Minister of Natural Resources and Tourism, Zimbabwe, 'Ivory Trade: The Zimbabwe Position', Department of National Parks and Wildlife Management, Harare, Zimbabwe, September 22, 1989, press conference, 2–3.
38 CITES, Fourteenth Annual Report of the Secretariat, Lausanne, Switzerland, January 1–December 31, 1989, 13.
39 CITES Secretariat, 'Interpretation and Implementation', document 7.23, October 9–20, 1989.
40 TRAFFIC-USA newsletter, 10 (1), March 1990, 9–10.
41 CITES has a 'reservation' clause allowing parties to exempt themselves from trade controls on any species listed in the appendixes. Reasons for taking reservations need not be given. Parties taking a reservation on a species are treated as non-parties for all issues regarding that species. In 1985, fifteen of the then eighty-seven parties had such reservations in effect; Japan led all others, with thirteen. Simon Lyster, *International Wildlife Law* (Cambridge: Grotius Publications, 1985) 9–10, 262–3.
42 TRAFFIC-USA newsletter, 10 (1) March 1990, 17.
43 US Department of State, telegram from American Embassy, Gaborone, Botswana, to Department of State, February 12, 1990 [R 121326Z Feb 90].
44 TRAFFIC-USA newsletter, 10 (1) March 1990, 17.
45 Allan Thornton, 'A Ban on Ivory to Save the Elephant?', *New Scientist*, September 30, 1990, 43.
46 Nick Cater, 'Preserving the Pachyderm', *Africa Report* November/December 1989, 45.
47 Thornton, 'Ban on Ivory', 43.
48 Michael A. O'Connell and Michael Sutton in cooperation with TRAFFIC-US, *The Effects of Trade Moratoria on International Commerce in African Elephant Ivory: A Preliminary Report* (Washington, DC: World Wildlife Fund and the Conservation Foundation, 1990), 18.
49 US Department of State, telegram, American Embassy, Abu Dhabi, UAE to Department of State, May 16, 1990.
50 O'Connell and Sutton, *Effects of Trade Moratoria,* 12–13.
51 Ibid., 26–7.
52 Ibid., 28–30.
53 Dr I. Douglas-Hamilton, WWF Project 3882, EEC/WWF African Elephant Programme, Activities Report, Nairobi, Kenya, reporting period: February 15, 1989–March 30, 1990.
54 O'Connell and Sutton, *Effects of Trade Moratoria,* 32.

55 For the moment, I ignore the extreme animal rights perspective, which holds that the species should never be exploited in the first place and a ban thus corrects this wrong and must be permanent. Here, I take a strict resource conservation perspective and, later, in the specific context of competing NGO perspectives, bring in the countervailing view. Consequently, I assume that a ban is only temporary, that it offers a breathing spell, and that some level of legal exploitation will resume.

56 For the classic statement on focal points in a bargaining and coordinated games context, see Thomas C. Schelling, *The Strategy of Conflict* (Cambridge, MA: Harvard University Press, 1960; reprint 1980).

57 In the abstract, imagine there are 1 million elephants with marketable tusks. At any time, anywhere from zero to two million tusks (two per elephant) could enter the market. Some number between zero and two million results in sustainable harvests. But for biological, physical, financial, and political reasons, no one can agree what that number is. No claim to the 'right' number has prominence. The only prominent number for a sustainable level is zero.

58 See, for example, Michael Lipsky, *Street-Level Bureaucracy: Dilemmas of the Individual in Public Services* (New York: Russell Sage Foundation, 1980).

59 Edward B. Barbier, Joanne C. Burgess, Timothy M. Swanson, and David W. Pearce, *Elephants, Economics and Ivory,* (London: Earthscan Publications, 1990), 2.

60 Notice that the implicit dependent variable in this discussion is *change*, not balance of power, world order or even international cooperation. This emphasis is itself a considerable departure from prevailing theories of international relations, including those in the regime and cooperation literatures. It follows from a central premise of this book – urgent ecological crisis.

61 For discussions of the collective action problem, see Russell Hardin, *Collective Action* (Baltimore, MD: Johns Hopkins University Press, 1982); Mancur Olson, *The Logic of Collective Action: Public Goods and the Theory of Groups* (Cambridge, MA: Harvard University Press, 1965).

62 On the targeting of foreign elites by moral entrepreneurs, see Ethan A. Nadelmann, 'Global Prohibition Regimes: The Evolution of Norms in International Society', *International Organization* 44(4): 479–526. Where elite behaviour cannot be targeted, one can expect environmental tactics like the ban to be less effective. In fact, this points up the strategic, let alone ethical, difficulty of targeting change efforts at those who acquire subsistence from the activity – for example, slash and burn farming – and those who acquire luxury benefits from that activity – for example, throwaway chopsticks.

63 Of course, it stretches credulity to claim that this one victory for the elephant will save the biosphere. It is possible, however, to conceptualize some threshold number of similar victories so that significant numbers of consumers eschew all products that are, or are suspected as being, tainted with ecological destruction.

64 A body of law and administrative machinery is an imperfect measure of a regime's success. Little, if any, research has been done on the extent to which any CITES-protected species has actually ceased to decline or has improved because of CITES implementation. See Laura H. Kosloff and Mark C. Trexler, 'The Convention on International Trade in Endangered Species: No Carrot, But Where's the Stick?', *Environmental Law Reporter* 17 (July, 1987: 10228). Assuming the lack of such evidence is the result of uncertain biology, the complexities of trade (e.g., multiple and shifting ports of entry, ease of concealment) and the costs of such research, not to mention the low priority governments attach to such issues, the ban has the advantage of being easily measured. When, for example, WWF wanted to ascertain the impact of the ivory ban, it merely called a half-dozen ivory retail outlets and discovered that the US market had dried up (Michael O'Connell, WWF, interview, 1991).

65 The ban as a core ingredient in an evolving, broad-based, biodiversity regime is consistent with the increasingly used international environmental norm known as the precautionary principle. According to this principle, scientific uncertainty of an environmental risk is not enough to delay action. In the 1970s proposals for a moratorium on whaling were justified on the basis of uncertain population impacts, rather than on the basis of scientific evidence that whales were threatened with extinction. (See Daniel Bodansky, 'Scientific Uncertainty and the Precautionary Principle', *Environment* 33(7) 1991: 4–5, 43–4; Kevin Stairs and Peter Taylor, 'Non-Governmental Organizations and Legal Protection of the Oceans: A Case Study' in Andrew Hurrell and Benedict Kingsbury, eds, *The International Politics of the Environment: Actors, Interests, and Institutions* (Oxford: Oxford University Press, 1992) 110–41. Similarly, despite the rapid decline in elephant populations, few observers were predicting extinction, even as late as 1989. The ban, in effect, said that we may not know until it is too late if the elephant is on the verge of extinction, that the physical, biological, and social aspects of maintaining habitat and genetic diversity are too uncertain, but that action, nevertheless, was necessary.

66 To illustrate the emphasis on enforcement, the otherwise excellent article by Kosloff and Trexler concludes:

> CITES is a legal measure where, due to the many conceptual and actual implementation difficulties involved, success needs to be proven rather than assumed. With other laws, a case study showing failure might be a surprise; with CITES, a case study showing success would be the surprise. The difficulties encountered in implementing CITES in the United States can only be magnified in other countries with fewer resources and less sophisticated infrastructure to dedicate to the issue.
>
> Kosloff and Trexler, 'International Trade in Endangered Species', 10236.

This is all true, but it misses the more nuanced effects of norm creation and the politics of non-state actors. This shortcoming highlights the usefulness of regime theory and much of the conflict resolution literature that attempts to account for the 'power' of non-state actors. On this last point see, for example, the analysis of Vatican and Quaker mediation in Thomas Princen, *Intermediaries in International Conflict* (Princeton, NJ: Princeton University Press, 1992).

67 Nadelmann, 'Global Prohibition Regimes'. By 'prohibition regime' I do not mean a total and permanent shutdown of exploitation and trade. Rather, a temporary ban may be enough to significantly reduce consumer demand in the long run, even when trade is resumed.

68 Ibid., 484–6.

69 John McCormick, *Reclaiming Paradise: The Global Environmental Movement* (Bloomington: Indiana University Press, 1989), 8.

70 Ibid., 17–18.

71 The following historical account is based largely on Robert Boardman, *International Organization and the Conservation of Nature* (Bloomington: Indiana University Press, 1981).

72 Quoted in Ibid., 37.

73 Quoted in Ibid., 42.

74 A few years later, Americans set up a US branch. Unlike most national affiliates, which contribute up to one-third of their funds to WWF International, the US group contributes none of its income. Nicholson explained in a recent interview that the American branch 'has always tried to go its own way. . . . The Americans didn't see the point in a global organisation or a world fund.' Fred Pearce, *Green Warriors: The People and the Politics Behind the Environmental Revolution* (London: Bodley Head, 1991) 7.

75 Interview in Pearce, *Green Warriors*, 7.
76 Ibid.
77 Boardman, *International Organization*, 80.
78 IUPN's trade monitoring arm, the Survival Service Commission, established in the 1940s, became formalized as TRAFFIC with headquarters in London. TRAFFIC acted as a data bank on trade in wildlife through a consultancy agreement concluded in 1978 with the CITES Secretariat. Boardman, *International Organization*, 93.
79 Nadelmann, 'Global Prohibition Regimes', 485.
80 One reason the number of parties has grown from ten at the convention's inception to 110 is that national environmental and trade NGOs pressure their governments to respond to declining populations. Because governments often have more pressing issues, joining the convention gives them access to information, funds, and technologies they would not otherwise generate on their own. Some parties, including the United States, will even condition bilateral technical assistance on joining the convention. In addition, many countries have nothing but CITES to protect species and habitat because their domestic laws and programmes are so ineffective. Susan Lieberman, US FWS, interview, 1991.
81 In extreme situations the secretariat has urged punitive actions against governments. For example, because Thailand has no domestic legislation implementing the convention and, thus, no science or management authorities, it has become a hub of illegal traffic. As a result, the secretariat has recommended a ban on all trade with Thailand by CITES Parties. Susan Lieberman, US FWS, interview, 1991.
82 CITES, Fourteenth Annual Report of the Secretariat, (1 January–31 December 1989); Lausanne, 16–17.
83 CITES, 'Interpretation and Implementation', 717.
84 Kosloff and Trexler, 'International Trade in Endangered Species', 10225–6, 42n. At the seventh meeting in 1989, TRAFFIC paid $10,000, WWF International $24,000, and various trade groups $75,000; all out of a total of $411,000 to a fund to help defer delegates' expenses. CITES *Proceedings*, October 8–20, 1989.
85 Kosloff and Trexler, 'International Trade in Endangered Species', 10226, 43n.
86 Ibid., 10226.
87 Susan Lieberman, US FWS, interview, 1991. At the 1989 meeting, the most prominent NGOs were WWF, TRAFFIC, Environmental Investigation Agency (EIA), Greenpeace, the Humane Society, and Safari Club International.
88 The International Whaling Commission does now have one NGO representative. Susan Lieberman, US FWS, interview, 1991.
89 Nadelmann, 'Global Prohibition Regimes', 482.
90 Ibid.
91 Pearce, *Green Warriors*. Because Pearce's book is devoid of references or even a bibliography, it is difficult to confirm his interpretations. He did, however, have access to many of the key British players, and it is on this basis that his work is cited.
92 This article and associated letters were included in a WWF-UK memorandum of April 4, 1990.
93 Pearce, *Green Warriors*.
94 For a critical assessment of this bandwagon effect, an assessment that fails to account for the international linkages and the provisional utility of a ban, see Raymond Bonner, 'Crying Wolf Over Elephants: How the International Wildlife Community Got Stampeded into Banning Ivory', *New York Times Magazine*, February 7, 1993, 16–19, 30, 52–3.
95 US FWS, interview, 1991.
96 For a more extensive analysis of the differential impacts of the ban on two African

range states, Kenya and Zimbabwe, with attention to the history of international conservation in Africa and the growing awareness of the necessity of local control of resources, see Thomas Princen, 'Ivory, Conservation, and Transnational Environmental Coalitions', in Thomas Risse-Kappen, *Bringing Transnational Relations Back In* (forthcoming).

97 Linking the local and the global is a fundamental analytic point to the thesis of this book. For amplification, see Chapter 2 and the conclusion, Chapter 8.

98 This proposition contains several qualifiers, each warranting much fuller development. 'Local empowerment', for example, is a controversial term because, in practice, it often translates into local co-optation by existing elites. For discussion with respect to international development and conservation, see L. David Brown and David C. Korten, 'Understanding Voluntary Organizations: Guidelines for Donors', WPS 258, (Working Paper, Country Economics Department, World Bank, Washington, DC, September, 1989); and with respect to northern environmentalism see M. Dowie, 'American Environmentalism: A Movement Courting Irrelevance', *World Policy Journal*, Winter 1991–1992; and Nicholas Hildyard, 'My Enemy's Enemies. . .' *The Ecologist* 23(2): 42–3, March/April 1993.

99 The 'must' can be interpreted in both a normative and a predictive sense. In the normative, I assume global society has an obligation to preserve biodiversity. In the predictive, if unsustainable practices continue, some set of consumers – a subset of current consumers or future generations – will live without that resource.

100 'Higher values' does not connote preservation, that is, abstinence from any exploitation. It does, however, imply that sustainable use is a higher value than short-term, unsustainable exploitation. In other words, sustainability, which allows for exploitation but not for irreversible damage, is consistent with the premises of this book.

101 Stephen Schneider, 'The Science/Policy Interface: How Much Knowledge Should Precede Action?', *Global Change*, Newsletter of the Project for the Integrated Study of Global Change, University of Michigan 1(2) 1991: 1–4.

102 For elaboration of the role of NGOs and social learning in the context of international development, see David C. Korten, *Getting to the 21st Century: Voluntary Action and the Global Agenda* (West Hartford, CT: Kumarian Press, 1990).

6 The Antarctic Environmental Protocol: NGOs in the protection of Antarctica

Margaret L. Clark

For six years, from 1982 to 1988, the members of the Antarctic Treaty System (ATS) negotiated guidelines for mining activities in Antarctica. But the final agreement, the Convention on the Regulation of Antarctic Mineral Resources Activities (CRAMRA), never came into force. Ratification required the signatures of all seven territorial claimant states, two of whom, Australia and France, shifted their influence and support to the formation of a World Park in the Antarctic and a complete ban on mineral activities.[1] On October 4, 1991, in Madrid, Spain, the Protocol on Environmental Protection to the Antarctic Treaty was signed banning all mining activities including prospecting, exploration, and development in the Antarctic region for fifty years.

Throughout this process, international environmental NGOs played an important role. The purpose of this chapter is to characterize this role and, in particular, to show how it evolved along with, and as a contribution to, the evolution of the Antarctic scientific and resource regime. I begin by setting NGO relations in the context of the ATS and CRAMRA and the overall Antarctic decision-making process. I then examine NGO attitudes and activities during the minerals negotiations and assess NGO impact on the policy shifts of Australia and France. Finally, I show how, in the four negotiating sessions resulting in the Environmental Protocol, NGOs were able to successfully promote the concept of a 'world park'. I conclude by arguing that, despite several features unique to Antarctica, this case holds general lessons regarding the role of NGOs in world politics.

THE ANTARCTIC TREATY SYSTEM

A United Nations General Assembly (UNGA) document describes the Antarctic Treaty as 'one of the most significant post-war contributions toward averting nuclear weapons proliferation and halting the nuclear arms race'.[2] The Antarctic Treaty System (ATS) has not been static, however. It has evolved to reflect the changing interests and concerns of the membership, including the effects of new technologies and, increasingly, of the global environmental crisis.

The ATS is a 'two-tiered' system in which Antarctic Treaty Contracting (or Consultative) Parties (ATCPs) have full voting rights and Acceding (or Non-consultative) Parties (NCPs) have no voting rights. Any state that is a member of the United Nations, or is invited to become one, can accede to the Antarctic Treaty (Article 13 [1]).[3] To acquire voting rights a state must conduct 'substantial scientific research there, such as the establishment of a scientific station or the despatch of a scientific expedition' (Art. 9 [2]). As will be seen, two NGOs have met this criterion (albeit, as non-state actors), and in the process, gained credibility both for their science and for their politics.

The two-tiered system also distinguishes between those states with territorial claims, the *claimants*, and those states without claims, *non-claimants*. Among the original twelve signatories to the Antarctic Treaty, seven are claimants: Argentina, Australia, the United Kingdom, Chile, France, New Zealand, and Norway. All the Antarctic claims remain legally 'frozen' by Article 4 of the treaty. Significantly, two other original signatories, the United States and the Soviet Union,[4] maintain an option to establish claims in the region.[5] The other original treaty members were Belgium, Japan, and South Africa.

Aside from the well-publicized 'non-militarized' and 'non-nuclear' provisions of the Antarctic Treaty (Art. 1,5), other provisions of the treaty were made to foster cooperation and transcend narrow national interests.[6] One such example is Article 7, which states that all Antarctic facilities, equipment, ships, and aircraft may be subject to inspection by any ATS participant at any time.[7] This 'right of inspection' includes aerial observation of all facilities. The norm associated with this right has effectively been extended to or, perhaps more correctly, assumed by, non-state actors such as NGOs, which monitor and report on state-sponsored activities. As will be seen, Greenpeace in particular has appropriated this norm in monitoring its own research station and those of member states for environmental impacts.

Another aspect of the ATS is the free exchange of scientific information and scientific field personnel, which has created bridges of understanding among the various cultures and political ideologies.[8] Logistically, the exchange of personnel has also enabled those field programmes without either base operations or strong financial support to function in Antarctica. In some cases these joint scientific operations and their findings have generated enthusiasm (and, therefore, financial support) among national decision-makers. These activities have led to the building of new field stations and the subsequent admission of new, full voting members to the Antarctic Treaty System. One such example was the cooperative field work conducted by American and Chinese scientists and the later establishment of the Great Wall field station by the People's Republic of China, which became a full treaty member in 1985.

The norm of open scientific exchange has also been appropriated by NGOs conducting their own research on the continent. By operating a station and conducting bona fide scientific research, Greenpeace, and more recently, the

Antarctic and Southern Ocean Coalition (ASOC), have participated in enhancing overall scientific understanding. At the same time, the knowledge these NGOs contribute, although legitimate in its own right on scientific grounds, is aimed at enhancing the understanding of *environmental* impacts for members and other actors, and thus, has explicit political content.

The cooperative sharing of scientific data among states has been at odds with the operations of resource-related industries. Some scientists have felt pressured by both governments and industries to withhold information about mineral and natural gas occurrences in order to secure a competitive edge. The withholding of scientific information has prompted accusations among treaty members that treaty provisions have been broken. The incompatibility of cooperative science and competitive economic exploitation appears to have been exploited by NGOs and other opponents to the minerals convention who have argued that a fundamental norm of the overall ATS regime is violated by commercial exploitation. This argument may have contributed to the collapse of the minerals regime and the eventual signing of the Environmental Protocol.

Another point of conflict in the ATS is the use of a consensus approach to decision-making, where deliberations on any topic can be lengthy. This process encourages treaty members to table the more controversial topics, such as territorial claims, on which consensus cannot be reached. The resolutions that do emerge from this process reflect a compromise of opinions, the implementation of which is frequently successful.

These features of the ATS have been challenged by developments in science and technology, especially the increased ability of economic actors to discover, acquire, and exploit Antarctic resources. The Antarctic Treaty members have modified the ATS by creating and attaching specific topic conventions or measures to the treaty. These are: the Agreed Measures for the Conservation of Antarctic Flora and Fauna (1964), the Convention for the Conservation of Antarctic Seals (1972), the Convention on the Conservation of Antarctic Marine Living Resources (CCAMLR), (1980) and, now, the stillborn Convention on the Regulation of Antarctic Mineral Resources (CRAMRA, 1988). Each measure reflects changes not only in science and technology but also in the perceived ability of member states to manage the region's resources and environment. It was in the CRAMRA negotiations that environmental NGOs began to seriously doubt this ability and thus became most involved in Antarctic issues.

INTERNATIONAL NGOs

Non-state actor participation in Antarctic policy-making began when the Antarctic Treaty was negotiated in the late 1950s. The treaty calls for the 'establishment of cooperative working relations with those Special Agencies of the United Nations and other international organizations having a scientific or technical interest in Antarctica.' At the time, the participation of scientific groups was seen as a way to manifest and practice the ideals of the United Nations. But as mineral exploitation and its expected environmental effects became a real possibility, environmental NGOs saw their area of expertise to be as vital to the Antarctic as that of the more traditional scientific NGOs.

The most prominent traditional Antarctic NGO has been the Scientific Committee on Antarctic Research (SCAR). SCAR actually predates the Antarctic Treaty; it replaced the Comité Special de l'Année Géophysique Internationale (CSAGI), which coordinated the scientific activities during the International Geophysical Year (1957–8). SCAR is an independent scientific advisory body which approves all scientific activities within the Antarctic Convergence. It 'provides a link between the Consultative Parties and national scientific committees also offering a framework for the international cooperation of such projects.'[9] SCAR is composed of delegates from each of the national Antarctic research committees, one delegate from the International Council of Scientific Unions (ICSU), three delegates from ICSU-affiliated scientific unions, and one delegate from the World Meteorological Organization.[10] SCAR's bulletin is published three times a year by the Scott Polar Research Institute in Cambridge, and distributed throughout the scientific community. The various working groups of SCAR, each of which has an area of expertise such as biology or oceanography, provide scientific advice to the ATS members. SCAR also conducts environmental impact assessments (EIA) and recommends sites for protection.[11] One example was an airstrip built by France without informing anyone. Field programmes are supposed to submit any building or remodelling plans to SCAR for an EIA. Evidently no such report was filed. The first EIA on the airstrip was conducted during the building phase by Greenpeace.[12]

During the early phase of the negotiations over the minerals convention, SCAR proposed a conservation area for Antarctica, similar to the 'biosphere reserve' designation applied by UNESCO's Man and the Biosphere programme. SCAR also initiated an international ten-year study of the Antarctic biosystem, the Biological Investigations of Marine Antarctic Systems and Stocks (BIOMASS). Member governments, however, have not been willing to finance such a large operation.[13] The concept of considering the environment holistically when making resource policy decisions has become a trademark of SCAR's work. Remaining focused on scientific, not political issues, continues to be another SCAR trademark. As SCAR president J. H. Zumberge states: 'We have managed to keep SCAR clear of involvement in the international politics of Antarctica. We accomplish this mainly by concentrating on science and leaving the politics to the Consultative Parties.'[14] But if politics could be avoided in the first two decades of the ATS, it could not in the 1980s, when mineral exploitation became prominent. Furthermore, many environmental NGOs reasoned that neither the science nor the politics could be left to the official parties.

Thus, although SCAR has remained prominent on scientific grounds, the environmental NGOs have taken the lead on issues of protection and, in some cases, even research. In this capacity, they have enhanced environmental understanding among decision-makers, the scientific community, the global media, and the public at large. They have monitored scientific field operations and political negotiations alike, reporting their findings to the public. They have helped build coalitions among states and other interested parties, not only by providing information, but by sponsoring symposiums and conferences where Antarctic participants and others could meet. Most significantly, environmental NGOs have continued to focus attention on the norms of peaceful use and protection of the region embodied in the Antarctic Treaty. Greenpeace International, the Antarctic and Southern Ocean Coalition, and the International Union for the Conservation of Nature have been the most visibly active.

Greenpeace International has invested the greatest amount of resources in Antarctica, both financially and in human terms. The publishing division of Greenpeace produces glossy table-top displays of Antarctica with heart-tugging commentary about the need to protect the world's last wilderness. It also produces scientific fact-sheets. Public demonstrations by Greenpeace members in front of embassies are well known and have helped to keep Antarctic issues alive, both in the press and with the public. But it is the Greenpeace field station and the organization's first-hand knowledge of the region's difficult operational circumstances that have given this NGO credibility as a significant representative of the Antarctic environment.

The World Park Base was established in January 1987 at Cape Evans, Ross Island.[15] The year-round facility supported four people during the winter, more during the summer. An annual visit by the Greenpeace ship, MV *Gondwana* resupplied the station, exchanged personnel, and removed the refuse and human waste to New Zealand. The station used several forms of 'alternative living' to prove the viability of low-impact operations and to set an example for personnel at the other, state-run stations. Waste is an especially difficult problem because extreme Antarctic temperatures prevent human by-products from decomposing. Most stations either bury or throw into the Southern Ocean their waste materials, including human wastes. The Greenpeace staff, by contrast, used a composting toilet, which uses aerobic bacteria and needs no additional water or chemicals, thereby reducing waste disposal by 85 per cent.[16] The station also made use of wind and solar energy, substantially reducing their need for diesel generators, which most field stations rely on for power. Greenpeace also sent an annual 'Environmental Impact Assessment and Exchange of Information' report to all Antarctic Treaty members, both contracting members and non-contracting members.

Although few in number, the Greenpeace field staff undertook a variety of scientific projects. These included limnological studies of the lakes of Ross Island, zooplankton studies of Cape Evens lakes, astronomical observations, mapping of the glacial erratics on Ross Island, and paleontological studies of the Quaternary sediments on Ross Island. Such year-long scientific work provided data to a variety of scholars besides Greenpeace scientists. Greenpeace thus operated within the ATS norm of free and open scientific exchange. At the same time, such work furthered the organization's legitimacy as an actor that does more than just protest and lobby.

Aside from practising science, Greenpeace personnel monitored pollution at other stations. They visited a large number of stations, took thousands of field samples and photographs, and made public their findings, both positive and negative. These monitoring teams reminded station managers of their environmental obligations under the Antarctic Treaty System and left behind written reports of changes that would have to be made to comply with the ATS. Follow-up visits were timely and findings were reported. In the early 1990s, visits were made to the Great Wall base (PRC), Bellingshausen (Russia), Teniente Marsh (Chile), Marambio (Argentina), Dumont d'Urville (France), and McMurdo (US).

The base was closed following the 1992–3 field season, owing largely to financial constraints suffered by the organization. Even so, Greenpeace was able to set yet another example for other base operators. When a base is retired, the usual procedure is to simply abandon the facility. Greenpeace plans to conduct and make public an EIA on the area after the base is gone.

The second most prominent environmental NGO is the Antarctic and Southern Ocean Coalition (ASOC), formed in 1977. It has some 200 NGO members from forty-nine countries and it includes the American Cetacean Society, Animal Societies Federation of New South Wales (Australia), Canadian Nature Federation, Centro Científico Tropical (Central America), Deutsche Naturschutzring (Germany), Environmental Defence Fund, Friends of the Earth (from various countries), Greenpeace International, and Chikyu-no-Tomo (Japan).[17] The coalition has enabled a variety of NGOs to both participate in and to become knowledgeable of Antarctic issues without having to make a large investment in either personnel or money. As a result, small groups have been represented in Antarctic issues. Some groups that are part of the ASOC also act independently on Antarctic topics; Greenpeace International is one such example. The existence of a single coalitional organization also prevents like-minded individual groups from working against one another.

ASOC has had observer status at Special Consultative Meetings (SCM) where it regularly speaks out and circulates information. It has earned a reputation for collecting accurate information and for enthusiastically pressing for environmental protection of the region. ASOC is also responsible along with the Antarctica Project, Greenpeace International, World Wildlife Fund International, and the Cousteau Foundation, for the Antarctic edition of *ECO*, an 'occasional newsletter' published by Friends of the Earth International. *ECO* is published and distributed on-site during Antarctic-related negotiations. The purpose of *ECO* is to 'provide ideas and alternative proposals for benefits of delegates to intergovernmental meetings, and to clarify issues for the media.'[18] In addition, *ECO* staff members give radio and press interviews during the meetings. ASOC also disseminates Antarctic information through the 'Antarctica Project', an international citizen's network with a quarterly newsletter.[19]

ASOC, like many coalitional NGOs, does all of this with minimal organizational structure.[20] In fact, a single staff person in Washington, DC coordinates much of the operation with an informal secretariat comprised of Greenpeace Australia, ASOC/ECO New Zealand, and the Antarctica Project in Washington, DC.[21] Information and press releases are sent to ASOC member organizations, which are encouraged to provide feedback to ASOC. In 1991, ASOC was invited by Australia to observe their field operations on the Antarctic continent.[22] This arrangement makes ASOC the second NGO to either conduct or observe on-site Antarctic field operations and has proven to be a useful resource to both.

Two other NGOs, the International Institute of Environment and Development (IIED) and the International Union for the Conservation of Nature (IUCN), have taken both a traditional and an activist role in Antarctic politics. IIED has been involved in Antarctic matters since 1976. Representatives of this group have been regular attenders at Antarctic gatherings. Because other environmental NGOs supported a total mining activities ban, it was often IIED which provided the environmental conscience during the CRAMRA negotiations, pushing for high environmental standards should mining ever take place. During this time, an IIED representative often served as the 'public interest representative on US delegations'.[23] IIED 'Reports on Antarctica' were made publicly available throughout the CRAMRA negotiations.

IUCN has had a long history of Antarctic involvement. Along with the World Wildlife Fund (WWF), IUCN participated in the preparatory work for the convention on the Conservation of Antarctic Marine Living Resources. It has involved both WWF and IIED in joint projects. Such projects included a 1980 cosponsored workshop of 'scientific experts on krill fishing around Antarctica with the Center for Law and Social Policy and the Oceans Society.'[24] Another collective project that involved IUCN, WWF, and IIED examined krill and whale interaction to more accurately define a sustainable krill yield. By helping to develop both accurate baseline data and migratory information, these NGO projects determined environmental protection measures and sustainable yields for Antarctic fisheries. Such information was vital to the design of CCAMLR.

Although much of IUCN's work has fitted the traditional scientific and, hence, apolitical role of the early non-state actors, its work nevertheless contributed to the passage of the Environmental Protocol. In late 1990, IUCN wrote a strongly worded environmental document about Antarctica that called for an environmental regime and the exclusion of all mining. The document was sent to the Antarctic Treaties Contracting Parties (ATCPs) meeting at the Eleventh Antarctic Treaty Special Consultative Meeting, the first negotiating session, aimed at drafting an environmental protocol.[25] The IUCN document was again circulated during the second negotiating session, when the members accepted a version of the protocol. It is highly likely that the IUCN document influenced the work of the participants at both negotiating sessions.

In sum, Greenpeace and ASOC have carved out a niche in Antarctic politics that the more traditional NGOs were unable or unwilling to attempt. SCAR, IIED, IUCN and others remain active scientifically, but the on-the-ground politics of monitoring, reporting, and publicizing has largely fallen to these two NGOs. They have achieved a degree of legitimacy from their scientific research and their monitoring and from the example they set for low-impact operations that has enabled them to deal with states as more than mere pressure groups. Much as the coalitional NGO, Great Lakes United, adopted the water quality agreement as its own (see Chapter 4), these NGOs have adopted the norms of the ATS regime. The scientific NGOs also espoused the regime principles of scientific research and peaceful use, but Greenpeace and ASOC did more. As will be discussed shortly, they were able to help resolve the tension between ATS norms and minerals negotiations by promoting the concept of a World Park. Their persistence in developing and pushing this concept and their ability to back it up with concrete research and monitoring appears to account for their particular form of influence, especially in promoting the Environmental Protocol.

THE CONVENTION ON THE REGULATION OF ANTARCTIC MINERAL RESOURCE ACTIVITIES

The Antarctic Treaty nations spent six years, from 1982 to 1988, negotiating the Convention on the Regulation of Antarctic Minerals Resource Activities (CRAMRA). The entire exercise rested on two huge uncertainties: namely, whether there were indeed minerals on the continent, and whether the technologies would be developed to exploit them. Moreover, as noted already, unlike all previous measures negotiated by the ATS to modify the regime, this one challenged some fundamental norms of the regime itself. In the end, an elaborate and highly restrictive regulatory apparatus was not enough to mollify key actors, state and non-state.

Potential mineral resources

The geological reasons for mining in Antarctica are based on speculation. The Gondwanaland theory posits the break-up of an ancient supercontinent, at the centre of which was Antarctica. The known existence of offshore oil, natural gas, and manganese nodules (copper and nickel) on other parts of the former Gondwanaland has led to speculation that such resources exist off the Antarctic coasts. But the necessary technology for safely drilling and mining on either the Antarctic continent or off its shoreline has yet to be developed. Fixed platforms for drilling function well only in shallow waters (roughly fifty metres deep), as in those of the Arctic, but the Antarctic waters are much deeper.[26] Southern icebergs can also be huge; one calving created a 'bergie' the size of Rhode Island. The potential damage to a drilling platform in the path of such an iceberg would be costly, both to the environment and to an oil company's budget. The distance to markets and support infrastructure is much greater than from other sites, such as those in the Arctic. All told, these characteristics led one expert to speculate that 'a super-giant oil field would need to be found in Antarctica before there would be economic justification for proceeding to exploit it. There is no evidence that such a field or fields exist.'[27] As an oil industry representative once said to this author at an Antarctic conference: 'Antarctica is a hell of a place to wildcat.'

Minerals negotiations

If the existence of minerals is so uncertain, why were the CRAMRA negotiations initiated? There are three likely explanations. First, in a resource-hungry world, those states with the greatest technology and financial cap-abilities are likely to be the first on site and the first to exploit. Second, the discovery of marketable resources would be likely to melt the present legal 'freeze' on territorial claims. As a result, old territorial disputes would be renewed and the entire cooperative spirit of the ATS could end.[28] A third reason to initiate CRAMRA talks was the serious question of environmental impact regionally and globally. The influence of both the Antarctic continent and the Southern Ocean on global weather and marine life has become well documented, especially in the 1980s as discussed in Chapter 3.[29] Studies of Antarctic ice-core samples help scientists understand the world's historical climate.[30] Antarctic observation stations also monitor 'plasma disturbances to chart solar storms and improve worldwide weather forecasting.'[31] Infor-mation retrieved from the region enables scientists to better understand the world's weather, which, among other things, is valuable for agriculture worldwide. Another benefit of Antarctic research stems from the identifica-tion of the ozone hole over Antarctica's Halley Bay by the British Antarctic Survey in 1985. This discovery renewed discussion of the dangers of chlorofluorocarbons, a debate that had been stagnant since 1975.[32]

In short, if mining were to occur – which some believed likely to happen once price, demand and technologies existed – then the full implications of that mining had to be anticipated.[33] A minerals convention, consequently, was supposed to be as much a preventative measure as one to promote exploitation. Moreover, many actors reasoned that it would be easier to establish a stringent set of standards for mining before there was any serious intent to mine and before the distribution of the benefits of such mining could be established. In this regard, the uncertainties over resources and juris-dictions actually made it easier for all interested parties to put a high premium on environmental protection. CRAMRA depended in part on the negoti-ations' special features, especially the uncertainties regarding the amount and quality of the minerals, the lack of legal sovereignty over the area, and the fact that non-member states had a stake in the outcome because of its potential effects on the availability and price of energy.[34]

As for NGO positions, ASOC and Greenpeace maintained that nothing short of a complete ban on all mining activities would protect the environment. Groups such as IIED, as noted, felt that constructing a regulatory framework with very high environmental standards for all stages of mining activity, including the decision to prospect, was a more durable approach to environmental protection. These NGOs reasoned that, even with a moratorium in place, such as that which eventually happened with the fifty-year ban in the Environmental Protocol, CRAMRA safeguards and regulations were necessary in case such a moratorium was broken. As two legal scholars, Westermeyer and Joyner, note, 'Any moratorium agreed to – no matter its duration – can fall victim to a single nation's change of heart.'[35] As it turned out in the subsequent Environmental Protocol, a 'walkaway' clause allowing any country to leave the protocol after the fifty-year ban and still remain a member of the Antarctic Treaty, was included. Thus, in retrospect, support of CRAMRA regulations as a back-up mechanism did make sense, institutionally and environmentally.

CRAMRA was adopted by consensus on June 2, 1988 and opened for ratification on November 25, 1988. The agreement accommodated concerns between claimant nations and non-claimant nations, between the ATS nations and the international community, and between environmental considerations and the logistics of mining. Among the concessions to environmental NGOs in CRAMRA was the granting of observer status to NGOs. A second concession to NGOs was access to all non-confidential information and documentation. These two concessions were designed to enable NGOs to participate more effectively. Guidelines for gaining observer status varied among the newly created institutions. In the commission, for example, observer status was given primarily to those NGOs with the skills and knowledge to conduct environmental impact assessments. In the Special Meeting of States, organizations with observer status in either the commission or the Advisory Committee can gain observer status. The Advisory Committee bases its observer status on commission decisions. The Regulatory Committee has no specific provisions, but it does not prohibit NGO observer status.[36]

The one-year ratification process for CRAMRA held several surprises, due largely to position changes by Australia and France and, more generally, to growing acceptance of concepts such as 'common heritage' and 'world park'. In both respects, NGOs were prominent not just to 'educate' but to help the states resolve conflictual issues among themselves.

AUSTRALIA AND FRANCE CHANGE POSITIONS

Australia and France are both original signatories to the Antarctic Treaty. Based on a combination of early exploration and geographic proximity,

Australia claims 42 per cent of Antarctica, the largest of all claims. The French claim dates back to an 1840 exploration and is located between the two Australian claims. The French have focused on the rich fishing grounds of several subantarctic islands and the Terre Adélie shoreline.[37] Both Australia and France have active domestic environmental groups that have monitored their countries' activities in Antarctica. The personal influence of oceanographer Jacques Cousteau on the French government has been especially important. Cousteau's ability to speak directly to French policy-makers of his environmental concerns has kept these concerns politically viable.

France's failure to ratify CRAMRA can be traced to a decision in 1981 to level a chain of islands at Pointe Géologie and to fill the sea between them to create an 1,100 metre airstrip.[38] Construction began without either an environmental impact assessment or notification to the other treaty members. Greenpeace personnel took photographs showing the destruction of both the area itself and the breeding abilities of penguins and other seabirds that had rookeries at Pointe Géologie. Both domestic and international environmental groups were incensed. They were quick to identify the contradiction: France was agreeing to restrict mining activities in CRAMRA, while continuing to build the airstrip in violation of the ATS measures on the Protection of Flora and Fauna. In this context, France's decision to withdraw from CRAMRA and then to support the environmental protocol was probably an attempt to placate both the environmental groups and the general public, at a relatively low cost.

In Australia, several factors led to the government's rejection of the treaty, one of which was a highly visible lobbying campaign by national and international NGOs. Furthermore, the Australian Mining Industry Council (AMIC) was ambivalent about Antarctica. They always sent a representative to CRAMRA negotiations, as part of the Australian national delegation. Their position was that, while they had no plans to mine in the Antarctic region, they felt it was important for Australia to keep its mining options open and therefore it would be better to sign CRAMRA than not to. Australia's foreign and environment ministers both supported CRAMRA. They felt that it would provide the best chance to protect the Antarctic environment. Remembering the negative diplomatic reaction toward the US when the US failed to sign the Law of the Sea Treaty, the foreign ministry also expressed concern that Australia would face similar reactions if it did not sign CRAMRA.[39]

Both the Treasury Ministry and the Resources Ministry opposed CRAMRA. The treasury felt that signing the agreement would be 'tantamount to admitting that Australia does not "own" its own Antarctic territory'.[40] The Resources Ministry was concerned that a potential loophole in CRAMRA could allow 'states to subsidize unprofitable mining operations for strategic purposes', which would adversely affect the Australian mining industry.[41] The treasury also opposed the revenue-sharing scheme in CRAMRA. Under that scheme, when land was being considered as a mining site and when mining was under way, the relevant claimant state would sit on the specific regulatory committee and be expected to staff operations sufficiently to ensure all necessary rights and obligations. Given the size of the Australian claim, the government foresaw major staffing problems, because each mining bid would require the establishment of a separate regulatory committee. The financial return on this cost would be less than one-half of any revenues derived from marketing the resource. The Australian treasury felt that any amount short of one-half the financial return was unfair because the resource was non-renewable and the land would be permanently damaged.

The Environmental Minister, Graham Richardson, believed that mining in the Antarctic region was inevitable and, therefore, establishing regulations was environmentally responsible. His public statement that modern drilling and shipping practices negated any threat of a serious oil-spill in the Southern Ocean was followed, coincidentally, a few days later by the *Exxon Valdez* oil-spill (24 March 1989) in Prince William Sound, Alaska.

Prime Minister Hawke was under pressure from a variety of sources. Individual members of the Australian Senate lobbied each other both in support of, and in opposition to, CRAMRA. French oceanographer Jacques Cousteau wrote to the prime minister, 'urging Australia to take a lead in protecting Antarctica' and to reject CRAMRA. A visit from the French prime minister Rocard, specifically to discuss the French opposition to CRAMRA, helped persuade Prime Minister Hawke. At the same time, Australia's opposition party was publicly calling on the prime minister to take a stand against CRAMRA. Conservation groups met with Environmental Minister Richardson on May 12, 1989 and urged him to withdraw his support of CRAMRA.[42] On May 22, 1989, the Cabinet 'declared that it would not sign CRAMRA, a decision that took even the environmental groups by surprise'.[43]

Throughout this period, environmental groups in Australia took advantage of the government's accessibility to speak directly to policy-makers. Likewise NGOs were able to tap into a growing environmental conscience among both the Australian public and the policy-makers. Both Australian and New Zealand environmental organizations made Antarctica the centre of major public campaigns. Citizens in both countries are educated in school about Antarctica and have a sense of affection toward the region. Indeed, most Australians and New Zealanders consider Antarctica to be the world's last great wilderness and feel a strong sense of responsibility toward it. Unlike many northern audiences, they do not have to be educated by NGOs about where Antarctica is, its political and geological history, or its significance to science today; much of this is included in the school curriculum. Instead, NGOs can emphasize the importance of maintaining the pristine Antarctic environment, which is the major land mass in the southern hemisphere and just hours away from both Australia and New Zealand. This awareness opened the way for NGOs to launch both a media blitz and a public participation programme with postcards and other mass actions.

NGOs promote the world park concept

The sequence and timing of events in France and Australia and elsewhere (such as the *Exxon Valdez* oil-spill) and the pressure exerted on some governments by NGOs probably provide a sufficient explanation of the failure of the minerals convention. After all, each claimant state had an effective veto. What is more difficult to explain is the rapid and seemingly easy negotiations to conclude an environmental protocol. Part of the explanation is the ready availability of a concept that simultaneously preserved the norms of the ATS, helped member states avoid the intractable claims issue, and eliminated the entire mining question. The concept was a 'world park' status for Antarctica. Such status would be similar to that of a Man and the Biosphere's 'biosphere reserves'.[44] These reserves are 'intended to provide continuing protection for examples of the various biogeographic regions of the earth'.[45] It was also a concept that was ignored, even derided, throughout the minerals negotiations. But the NGOs kept the concept alive and, in fact, helped it become the conceptual basis of the Environmental Protocol. When circumstances were right, when the minerals convention collapsed and the spectre of the collapse of the entire Antarctic regime loomed large, a World Park concept fed perfectly into the ATS negotiations and to what would become the latest step in the evolution of the overall Antarctic regime.

The World Park concept can be traced to the notion of the 'common heritage of mankind' (CHM), which refers to responsible management of areas or resources not possessed by a state, or collective of states, and is thus the responsibility of the global community. The open seas, outer space, and the deep seabed are often identified as global 'commons' and, hence, subject to common heritage designation. Because the territorial claims of Antarctica are legally 'frozen', the region is sometimes perceived as a 'commons' as well. But it is a commons with access limited to those in the 'exclusive club' of the ATS whose interests may not coincide with global interests. Caldwell credits Dr Arvid Pardo, Malta's ambassador to the UN, for introducing the 'common heritage of mankind' concept on the 'agenda of world politics' in 1967 during the UNCLOS debates.[46] Significantly, the CHM leaves open the door for resource development and management, provided that any resulting wealth is used to foster the economic progress of the world's poorer regions.[47] In the case of Antarctic mining, environmental NGOs pointed out that while the world's poorer regions may gain financially (which was provided for in CRAMRA), they would also have to pay for any environmental costs. Throughout the world, poorer countries are very familiar with the environmental costs incurred through resource development. Antarctic mining could have far greater – and far more negative – implications.

Although the term 'common heritage of mankind' speaks to global responsibilities both today and in the future, the term 'world park' conjures the idea of a pristine and protected environment. Married together, as done by environmental NGOs, the concept becomes one of a protected Antarctic environment, held in trust by today's global community on behalf of future generations. Thus, the term 'world park' incorporates the notion of global interests and global responsibilities. It implies that all interested actors, state and non-state, organized or as individuals, present and future, have a right to see or, at least, to know, that Antarctica will remain as it is indefinitely. A world park designation also maintains both the spirit of international scientific cooperation and the use of the region for peaceful purposes.

According to Greenpeace International, the following principles would govern World Park Antarctica: 'to protect the wilderness and the wildlife, to maintain international scientific cooperation, and to maintain a zone of peace'. These ideals have been embodied in the Greenpeace 'Antarctic Declaration', a citizen petition that had acquired 1.5 million signatures from people representing seventy-nine states.

The concept of a world park did not really gain credence until the minerals convention died and the member states faced a predicament.[48] At this time, member states knew that, if minerals were discovered and extraction technologies developed, the rush to claim territorial and extraction rights would be likely to doom the entire ATS regime. They had to either renegotiate the entire minerals convention, which had already taken six arduous years, or replace it with something new. For most member states, the primary concern was the maintenance of the Antarctic regime, not minerals exploitation. Thus, a concept that was simultaneously consistent with regime principles and afforded the states a way out of their predicament would be most welcome. Here is where the persistence of the NGOs in promoting the world park concept during the minerals negotiations paid off. Their message was now heard in a new light, not because the member states were suddenly convinced of the environmental needs of complete protection, but because their primary concern was regime maintenance. The NGO message thus helped them out of this predicament. The world park concept was also attractive because many likely prohibitions under such a designation were already present in the 1959 Antarctic Treaty: no military activities, no nuclear activities for either energy or defence, and no disposal or storage of nuclear or toxic wastes. In addition, the killing of marine life and birds was already restricted by the CCAMLR agreement. The monitoring of other activities, such as tourism, commercial fishing, the operation of support facilities, and the building of new facilities was also stipulated by existing conventions.

In sum, given the failure of the minerals convention and the international community's increasing sensitivity generally to global environmental issues, a world park designation for the Antarctic was a logical step to take. Nevertheless, the practical application of the concept raised new issues for the ATS to resolve. Tourism was one.

Overflight tourism, flying over the continent while viewing scenic locations, rose to three thousand flights a year in 1990.[49] The majority of tourists visit the continent via ship tours, which must comply with guidelines that limit both where tourists can visit and how long they can stay on the continent. Even short visits, however, can have a devastating impact on the fragile Antarctic ecosystem. In emergencies, search and rescue operations for tour groups must be undertaken by personnel from field stations in the area, thus jeopardizing even more lives. There are two primary problems in both the establishment of guidelines for tourism and in monitoring actors' compliance. The first is to determine who is responsible for those tour groups that are not sponsored by ATS member countries and the second is to answer questions of national jurisdiction when a tourist from one ATS state breaks laws in the territorial claim 'belonging' to another ATS member. Recognizing such problems, the Antarctic Treaty Special Consultative Meeting on Tourism was held on November 9–11, 1992, in Venice. ASOC was an observer at this meeting, which ended in a stalemate as participants could not agree on whether a separate annex on tourism was necessary. ASOC took the position that most of the concerns were already addressed elsewhere in the Protocol or in the Antarctic Treaty itself.

A second issue in a 'world park' designation was the question of administrative responsibility. Some argued for UN administration under either an already established UN agency, such as the United Nations Environmental Programme, or the creation of a specialized Antarctic Environmental Protection Agency (UNAEPA). But this raised the question of who, then, would manage the UNAEPA. One option would be for the present members of the ATS, plus a few additional states and NGOs. A second option would enable the current ATS to administer a world park. This option, of course, is not popular among non-members.

Negotiating the environmental protocol

As provided in the Antarctic Treaty (Art. 9(1)), decisions are made by the ATCPs during either Special or Regular Consultative Meetings. Special Consultative Meetings (SCMs) are held to deal with specific issues. The number of sessions of an SCM reflects the complexity of the topic. The SCM on the admission of China and Uruguay as consultative members lasted one day, but the SCM to negotiate CRAMRA lasted six years (1982–8).

In November 1990, representatives of the ATCPs met in Vina del Mar, Chile to begin the eleventh SCM. This meeting was a follow-up on environmental protection measures and mineral liability questions which had been introduced by Treaty parties at the fifteenth Regular Consultative Meeting in 1989.[50]

ASOC was granted observer status and officially introduced a draft set of principles and objectives as a 'model Antarctic protection convention'.[51] Other groups of member states presented drafts for consideration. France introduced a 'four country proposal' (Australia, France, Belgium and Italy), a convention which called for environmental protection measures for both environmentally high- and low-risk activities. Mining activities would be in a 'prohibited' category. Several institutions would be established to carry out the protection guidelines and conduct environmental impact assessments.

The 'five country proposal' (Argentina, Norway, United Kingdom, Uruguay, and the USA) called for an environmental protocol with general principles and annexes on specific topics. It would create an Advisory Board and a Secretariat, but no enforcement and inspection body. Disputes would be resolved by mandatory and voluntary means. New Zealand's proposal involved strengthening all existing environmental measures and placing a permanent ban on mining. It would also establish an Inspectorate to provide the monitoring and assessment of Antarctic activities.

The structure of the SCM (following the plenary session) consisted of two working groups and simultaneous Heads of Delegations meetings. ASOC was a formal observer to all but the Heads of Delegations meetings. The working draft of the protocol which emerged from this SCM was a collaborative effort. It included a prohibition on any mining activities. It also included draft articles on environmental principles, scientific cooperation, environmental impact assessments, compliance and inspection measures, response actions and liability reporting by parties.

At the second session in Madrid in April 1991, work continued on the draft protocol. By the time the meeting began, several more countries had joined France and Australia (supported by Belgium and Italy) in calling for a permanent minerals activities ban and increased environmental protection of the region. The United Kingdom supported such a ban for a fixed number of years, with reviews every five years. Just prior to this second session, Germany announced support of a permanent ban and during the meeting Japan added its support.

The draft protocol which was completed during the second session identified specific priorities and guidelines for activities in the region. Antarctica was designated as a 'natural reserve, devoted to peace and science' (Art. 1), mining activities (except for scientific research) were banned (Art. 6), EIAs were required for all proposed activities (Art. 7), a new advisory committee was called for to provide scientific and environmental assistance to ATS members (Art. 10), and guidelines were set to modify the Protocol, but not the annexes (Art. 24). Under Article 24, the Protocol (including Article 6) can be modified during the first fifty years only by consensus of all Antarctic Treaty members (ATCPs and NCPs). After that period, 'any ATCP can call for a conference to review the Protocol. . . a decision to modify or amend the Protocol would be adopted by a majority of all parties, including a majority of ATCPs.'[52] For an amendment to come into force, three-quarters of the ATCPs, including all ATCPs at the time the Protocol came into force, would need to ratify the amendment. A special clause within Article 24 pertains to the amending of Article 6 (the prohibition of non-scientific mining activities). This clause states that the prohibition would remain until 'the ATCPs have negotiated a binding legal regime on mineral resource activities that assesses if any, and under what circumstances, such activities should occur.'[53] Any proposed amendment would require the consent of all the original signatories to the Protocol, thereby maintaining the ATCPs' veto power.

As for NGO participation during this session, individuals from environmental NGOs served on national delegations as advisors. They came from ASOC and Greenpeace (Australia), Greenpeace (Denmark), Wilderness Society (USA), CODEFF (Chile), ASOC (New Zealand), and the Cousteau Society (France). Some NGOs had observer status during the various meetings and workshops, although they were excluded from the Heads of Delegations Meetings. These groups were ASOC, SCAR, IUCN, WMO, the Commission of the EC and the Intergovernmental Oceanographic Commission. Many NGOs also monitored events from outside the meeting. These groups included the Antarctica Project, Greenpeace Spain, ASOC New Zealand, Worldwide Fund for Nature, Cousteau Society USA, and Greenpeace International. NGOs thus represented environmental concerns both within the SCM and outside of it as monitors. The NGOs circulated various documents and special reports, including the *ECO* newsletter. The Antarctica Project's newsletter did not appear until spring 1992, although their reports and information sheets were already publicly available. The groups also met with the media and sent press releases to newspapers worldwide which provided an additional source of information to interested parties and the public.

At the third session in June 1991, both ATS members and NGO representatives hoped to finalize and sign the Environmental Protocol. Taking no chances, the NGOs came out in force. There were problems for them, however, not the least of which was obtaining official invitations. The Spanish hosts did not send official invitations to those observers who had attended the previous session; the reasons behind the oversight are unclear. Some observers felt that the Spanish hosts intended the initial invitations to suffice for all sessions. Others felt that the oversight was intended to prevent unwanted parties, especially NGOs, from participating. Despite the lack of formal invitations, however, ASOC, IUCN, and the Intergovernmental Oceans Commission did attend. They were admitted as observers to all meetings except those of heads of delegations. Other NGO environmental groups that monitored the session included Greenpeace International, the Antarctic Project, World Wildlife Fund-UK, the Cousteau Foundation, and Greenpeace Spain. Once again, individuals representing NGOs advised national delegations.

With the exception of Article 24 (which set guidelines for amending the Protocol), in this third session members adopted all the Protocol articles, all four annexes and an interim report. The national delegations were small compared to their previous numbers at other sessions, the primary reason being that no one expected any serious debate. Consequently, there were few senior negotiators or diplomats on the national delegations. They had planned to arrive for the signing of the protocol anticipated later in the week. They and most observers were thus taken aback when the USA moved to change Article 24.

Just before the session began, the USA announced its position: the amendment procedure in the Environmental Protocol as it dealt with mineral resource activities made it impossible for any actor to ever lift the ban. The US was willing to accept the initial fifty-year 'ban' (so called, because a lift of mineral activities required consent by *all* ATS members), but it proposed an Article 25 which would allow any state to 'walk away' from the ban provision if 'ratification of an amendment was not achieved within three years.'[54] This meant that the only form of prohibition left would be the 51 per cent of any ATCPs (not all of the present ATCPs) required to accept an amendment.

The proposed clause did appear to contradict the goals of recent United States legislation, the Antarctic Protection Act of 1990, which states that 'It is unlawful for U.S. nationals to engage in or provide assistance to any Antarctic mineral resource activity, pending the entry into force of such an international agreement, which would provide an indefinite ban on Antarctic mineral resource activities.'[55] At the time President Bush signed this bill into law, the ratification of CRAMRA appeared likely. Because of the shift in position by Australia and France, however, and the imminent replacement of CRAMRA with the Environmental Protocol, the Bush administration was now attempting to create a means by which mineral activity could some day be initiated, hence the insistence on the 'walkaway' clause.

When the US delegation said it would not sign the protocol unless the proposed 'walkaway' clause was included, reactions were swift and angry. The story was prominently displayed in newspaper headlines and news agency reports. Examples from newspapers include: 'US Under Fire for Antarctic Decision' (*Canberra Times*, 22 June 1991); 'US backdown on Antarctic mining ban condemned' (*The Melbourne Age* (Australia), 24 June 1991); and 'US Compromise "Disastrous"' (*Weekend Australia*, 22–3 June 1991). In London, Greenpeace staged a demonstration at the US Embassy, while in Australia the Conservation Foundation targetted the US Embassy there. Similar protests occurred around the world. Australian Prime Minister Hawke wrote to President Bush, 'urging him to accept the ban' and said that 'Australia is disappointed the US couldn't agree to the proposal compromise.'[56] From an international conference in Tokyo, Japan, letters supporting the ban were sent to President Bush from legislators representing Japan, the United States, the European Community, and the former Soviet Union.

In the end, the US achieved the walkaway clause. After hours of debate in both informal settings and in small groups of key countries, a compromise text of Article 24 was reached. For the first fifty years after the Protocol comes into force the ban can be lifted only by agreement of all ATCPs. After that time, the ban may be lifted if agreed to by three-quarters of ATCPs and then ratified by three-quarters of ATCPs including all twenty-six of the current ATCPs. However, any nation can walk away from 'the provisions of the Treaty [and thus presumably to mine without regulation] if the amendment is not ratified within five years.'[57]

President Bush agreed to this measure on July 3, 1991 and twenty-three of the ATCPs formally signed the Antarctic Environmental Protocol on

October 4, 1991 at a special ceremony in Madrid. The remaining three ATCPs, India, Japan, and South Korea did eventually sign the protocol. During a Regular Consultative Meeting in Bonn, in October 1991, the parties agreed to ratify the protocol as quickly as possible and 'to apply its regulation in the interim as far as "practicable and feasible".'

Ratification by all signatories was expected within the following two years. By the summer of 1992, however, only Spain had ratified the protocol. In the United States, the Senate consented to ratification but 'recommended that the Protocol not become legally binding until implementing legislation was enacted for U.S. citizens.'[58] Several bills were introduced in the United States Congress in 1992[59] and, by late 1993, the bills were still sitting in committees.

As of late 1993, NGOs continued to lobby members of Congress, as well as the White House, to hasten the hearing process. Because Vice-President Gore visited Antarctica in 1988, NGO representatives hope to use this connection to the President to their benefit. They are also monitoring legislation which may be influenced by ratification of the Environmental Protocol.

The environmental protocol

The Protocol on Environmental Protection to the Antarctic Treaty consists of the protocol and four annexes, each dealing with a specific topic: Annex 1, environmental impact assessments; Annex 2, the conservation of Antarctic fauna and flora; Annex 3, waste disposal and waste management; and Annex 4, the prevention of marine pollution. Because the Antarctic Treaty was negotiated on the primacy of the internationalism of science and resource conservation, the primary objective of the protocol is to provide 'comprehensive protection of the Antarctic environment and the dependent and associated ecosystems and hereby designate Antarctica as a natural reserve, devoted to peace and science' (Art. 2). The means to accomplish these goals are set out in Article 3 of the protocol. It requires the regular and effective monitoring of all activities, including science, tourism, governmental, and NGO activities. Any such activities shall be 'modified, suspended or canceled if they result in or threaten to result in impacts upon the Antarctic environment or dependent or associated ecosystems' (Art. 3 [4.b]). The prohibition of mining activities is specified in Article 7.

Established also in the protocol is the Committee for Environmental Protection, which 'provides advice and formulates recommendations in connection with the implementation of this protocol for consideration at Antarctic Treaty Consultative Meetings' (Art. 12). Roles for NGOs on this committee are established in Article 2, which designates the committee's membership. NGO involvement comes from two sources. First, each member to the protocol has a representative on the committee who may be accompanied by 'experts and advisers' (Art. 11 [2]). Second, the committee shall invite to the sessions as observers 'relevant scientific, environmental or technical organizations which can contribute to its work' (Art. 11 [4]). In both cases, there are opportunities for NGOs to observe committee activities and to contribute to the discussions their expertise of the Antarctic environment. NGOs will also be able to keep the media and interested activists informed of both the committee's performance and the degree of compliance by the members.

CONCLUSION

On first examination, Antarctic politics may appear, for geographic, bio-physical and social reasons, entirely unique and thus to hold few general lessons. The region has no indigenous human population and, consequently, no native culture to be compromised, nor any native land-use rights to negotiate. The treaty was negotiated on the basis of the 'internationalism of science and the recognition that nature is a resource to be conserved.'[60] Putting these interests first is unlike any other treaty governing a major geographic region. Consequently, until the minerals negotiations, the emphasis on science and the relative political neutrality of the scientific enterprise has enabled Antarctic policy-makers to avoid much of the politics found elsewhere. In the ATS, decision-making is by consensus, not by voting. This method helps create a treaty system that is both durable and respected by its membership. In such a process, each consultative member has equal weight and a vested interest in participating in both the creation and the implementation of policies. As a result, the membership indeed has more of a 'club' feeling than other intergovernmental organizations or regimes. Consensus decision-making, however, is also slow and tends to lead to 'least common denominator' solutions.

Despite these apparently unique features, it turns out that on both environmental and political grounds, this case has several features in common with other regional and global issues. Most generally, the region has the characteristics of a common property resource, as do so many marine, aquatic, and atmospheric resources. Whether decision-makers live there or not, whether there is a 'local' in the common sense of the term, the relevant actors still must decide resource use questions on the basis of its resource characteristics. With respect to the commons, then, they must overcome free-rider problems and must devise assessment, monitoring and protective measures. And, much as in the Great Lakes case of Chapter 4 and the ivory case of Chapter 5, the relevant actors are not just the states sharing the resource. They include a wide range of actors, state and non-state.

The Antarctic case also shares with others the importance of the debates and interactions among scientists, policy-makers, and environmentalists. Moreover, as technologies and demand for resources change, it was apparent in this case that science was not enough – certainly not to resolve the minerals and environmental issues. It is here, as in so many cases, where politics and new political actors such as NGOs enter. These features of the science/policy interface are common to much of environmental diplomacy and require continued study and understanding.[61]

These general features suggest that the kind of politics evidenced here may be relevant elsewhere. In particular, as an instance of the evolving role of environmental NGOs in world politics, this case shows how NGOs interact with the state system, on the one hand, and how the NGO community itself is changing, on the other.

With respect to state interactions, the NGOs in this case clearly expanded their activities beyond the traditional activist roles of public demonstrations and public education. Although they committed considerable resources, holding rallies and informing decision-makers and the general public via publications and 'hot line' fact sheets, their real contributions were greater. One was conceptual, the other practical. In both respects, NGOs played a prominent role in drawing out the political implications of environmental change, actual or potential, in the Antarctic. And they operated simultaneously at the national and international levels. Their biggest challenge, and ultimate success, was in promoting the concept of the world park.

Throughout the minerals negotiations, the case for a world park could be made only on limited and, often, resented, preservationist grounds: it would be a huge, frozen Yosemite. As the environmental consequences of human activities in the Antarctic and elsewhere became better known and more salient politically – due largely to the efforts of politically active environmental NGOs – the world park concept became increasingly palatable. But until early 1990, a mineral convention was widely viewed as necessary and inevitable. The idea of banning mining entirely and establishing a world park was preposterous. But as this case study has shown, the persistence of NGOs in promoting the idea of a world park and their simultaneous coordinated work at the national and international levels, helped make the idea a reality. Had the NGOs not persisted with this concept through the minerals negotiations and ratification processes, it is likely that the apparent inevitability of minerals exploitation would have proceeded. That is, until the involvement of the environmental NGOs, with their stress on both science *and* politics, the minerals question was framed by states and traditional, scientific NGOs as a technical issue: exploration and exploitation was inevitable, so the best the regime can do is gather the best data and adopt the best technologies to protect the region. A politically informed view, by contrast, pitted the minerals convention against the larger Antarctic scientific and resource regime. So viewed, it was apparent to at least two member states that the two could not coexist.

The NGOs in this case thus served as key agents of change, defining and disseminating ecological concepts. They did not create all the conditions for regime change, of course, but they did provide a key conceptual ingredient, one consistent with existing institutional norms. Their ability to promote such a concept depended largely on their legitimacy, especially a legitimacy based not on protest and constituent pressure, but legitimacy based on their active and credible participation in the regime itself.

Greenpeace, by establishing its own, bona fide scientific research station, went as far as any non-state actor could (and further than most states could) toward meeting the requirements for membership in the 'club'. Even then, they went a step further. They did not just call for environmentally sound research practices or for monitoring. They actually implemented such practices and monitored others' stations. They thus gained their legitimacy in terms of the Antarctic regime's own norms and procedures. By itself, this activity reinforced and strengthened the existing regime by helping ensure the parties' compliance. But during the minerals convention and environmental protocol negotiations, this activity made more salient the importance and practicability of environmental protection. These NGOs, in a sense, put into practice an Antarctic world park well before the states took the notion seriously. These NGOs thus occupied and helped create a new niche in Antarctic politics, a niche that became somewhat institutionalized in the environmental protocol. In this respect, the NGOs did indeed lead and the states eventually followed.

With regard to the evolution of NGOs themselves, this case illustrates an increasingly common shift from the strictly (or self-avowedly) scientific to the scientific *and* political.[62] Some of the traditional scientific NGOs did adopt more politically active stances. But Greenpeace, ASOC, and others effectively supplanted these groups. The environmental NGOs conducted research and disseminated their findings and they went that one political step further by monitoring and exposing member states for violations of the rules the member states themselves devised. The scientific NGO will continue to play an important role in the Antarctic. But to the extent that issues arise with *political* dimensions, the niche carved out by the environmental NGOs will be critical.

The NGO community, as discussed in Chapter 1, is highly diverse and, as a result, the potential for conflict within the community is great. The Antarctica experience suggests, however, that these differences can be overcome. Thus, a second area of development within the NGO community appears to be an enhanced ability to resolve conflicts among NGOs. With NGOs involved in both the minerals convention and the environmental protocol negotiations, they did not waste their efforts discrediting the reputation of opposing NGOs. NGOs did oppose each others' positions regarding, for example, the merits of a strong minerals convention versus an outright ban. But the debates did not degenerate into attacks on individuals or organizations, as has happened in international conservation (Chapter 5), the UNCED process (Chapter 7), and elsewhere. In fact, over the years some individuals moved from one NGO to another. Interviews with NGO representatives who have been involved in Antarctic matters for years reveal that some still speak highly of former colleagues in other organizations.

This high degree of cooperation among NGOs may relate to the nature of the issues, not to mention the history of interstate cooperation. When the mission is straightforward – promotion of environmental protection, whether in a minerals convention or in an environmental protocol – and little home turf is at stake, these differences can be overcome. Under other conditions where, for example, the livelihood or moral sensibilities of constituents are at stake, such cooperation may not be so forthcoming.

Three general propositions can be derived from these observations. First, for reasons discussed in Part I of this book, NGOs can venture into international politics in ways which are closed to state actors. They can represent collective concerns that span state borders and socio-economic differences. Not distracted by the need to launch re-election campaigns, NGOs can remain focused on the issues and develop their own expertise and contacts. With respect to the Antarctic especially, they can adopt an interstate regime, act to strengthen its norms and procedures, and take measures to help change the regime as conditions warrant. In short, they can be significant political actors, devising and employing their own bargaining assets to promote environmental values.

Second, the activities of NGOs in representing Antarctica demonstrate how valuable NGO participation can be to those interests that would otherwise be marginalized. Thus, in a region that has no indigenous human population or a population that is politically weak and underrepresented, NGOs will be critical to expand the interests represented from the merely economic (or even scientific) to the full range of environmental issues including biophysical, cultural, and political.

Finally, the successful adoption of the world park concept suggests that, with the right idea, waiting can pay off. It is likely that, in most cases, the conditions for adoption of a new idea cannot be anticipated and NGOs cannot force them. But developing a concept and putting it into practice (however imperfectly as with the research stations and the monitoring) is a significant step toward institutional change and overall social learning.

NOTES

1 The territorial claims are legally 'frozen' by the Antarctic Treaty. Countries making such claims are Argentina, United Kingdom, Chile, Australia, France, Norway, and New Zealand. Ratification of the minerals regime also required the signatures of the United States, the former Soviet Union, five developing states, and eleven developed states.

2 UNGA A/39/583 (pt. 1), 1984, 46 in Beck, P. *The International Politics of Antarctica* (New York: St Martins Press, 1986), 314.

3 Article 13 (1) states:

> the present Treaty shall be subject to ratification by the Signatory States. It shall be open for accession by any State which is a Member of the United Nations, or by any other State which may be invited to accede to the Treaty with the consent of all contracting Parties whose representatives are entitled to participate in the meetings provided for under Article 9 of the Treaty.

4 For purposes of this chapter the term 'Soviet Union' or 'USSR' will be used to identify the political structure known as the Commonwealth of Independent States, although Russia has formally taken over the former Soviet Antarctic programme.

5 The significance of both the USA and USSR maintaining the option of establishing territorial claims stems from two factors. One was the politics of the Cold War and the perceived potential of Antarctica's geo-strategic location for both communication and transit needs. The second factor was that both states had made preliminary claims. In the case of the US, 'sixty-eight claim markers were dropped' from aircraft during Operation Highjump (1946–7) which determined boundaries, should the need arise. The Soviet claim was alleged to also be large. Shapley, D. *The Seventh Continent: Antarctica in a Resource Age* (Washington, DC: Resources for the Future, 1985) 52.

6 Article 1 (1) states:

> Antarctica shall be used for peaceful purposes only. There shall be prohibited, *inter alia*, any measures of a military nature, such as the establishment of military bases and fortifications, the carrying out of military maneuvers, as well as the testing of any type of weapons.

Article 5 (1) states that 'Any nuclear explosions in Antarctica and the disposal there of radioactive waste material shall be prohibited.'

7 Article 7 (2) states that 'Each observer shall have complete freedom of access at any time to any or all areas of Antarctica.'

Article 7 (3) states that 'All areas of Antarctica, including all stations, installations and equipment within those areas, and all ships and aircraft at points of discharging or embarking cargoes or personnel in Antarctica, shall be open at all time to inspection by any observers.'

Article 7 (4) states that 'Aerial observation may be carried out at any time over any or all areas of Antarctica by any of the Contracting Parties.'

8 Article 3 (1) states that 'information regarding plans for scientific programs in Antarctica shall be exchanged', 'scientific personnel shall be exchanged in Antarctica between expeditions and stations', and 'scientific observations and results from Antarctica shall be exchanged and made freely available'.

9 Beck, *International Politics of Antarctica*, 162.

10 Ibid. 162.

11 SPAs are Specially Protected Areas and SSSIs are Sites of Special Scientific Interest. The designation of SPAs and SSSIs are not always respected by field base operations. Penguin rookeries are often recognized as SPAs, and huts of the early polar explorers are designated as SSSIs.

12 This incident suggests one of the problems of having in-house inspection and monitoring mechanisms and seems likely to be a problem under the Environmental Protocol.

13 Funding for SCAR, which comes from member states, has always been limited. The 1983 Antarctic Treaty Consultative Member Meeting urged increased funding for the organization.

14 Beck, *International Politics of Antarctica*, 164.

15 The field station is located at 77° 38' South, 166° East, 25 km from the McMurdo Station (US) and the Scott Station (NZ).

16 Greenpeace Fact Sheet, 1991.

17 ASOC press release, June 14, 1991.

18 *ECO* 80 (2), April 22–30,1991, Madrid.

19 Barnes, J., *Let's Save Antarctica*! (Victoria, Australia: Greenhouse Publications, 1982).

20 Great Lakes United similarly has a small permanent staff, given its geographic and representational range as a coalitional NGO. See Chapter 4.

21 Kimball, L., *IIED Reports on Antarctica*, (Washington, DC: International Institute for Environment and Development, 1984–7).

22 Telephone interview with the Antarctica Project Director, Beth Marks, 20 September 1993.
23 Speech given by IIED representative Lee Kimball to the Antarctican Society, Washington, DC on 19 March 1985.
24 J. Barnes, *Let's Save Antarctica!* (Victoria, Australia: Greenhouse Publications, 1982).
25 Antarctic and Southern Ocean Coalition, press releases: April 30, 1991, June 14, 1991.
26 Barbara Mitchell and Jon Tinker, *Antarctica and its Resources* (London: Earthscan Publication: International Institute for Environment and Development, 1980).
27 C. Beeby. 'The Antarctic Treaty System as a Resource Management Mechanism –Nonliving Resource', In *Antarctic Treaty System: An Assessment*, ed. Polar Research Board, 1986, 272.
28 William Westermeyer and Christopher Joyner, *Negotiating a Minerals Regime for Antarctica* (Pittsburgh, PA: Pittsburgh University Press, 1989).
29 Patel, J., and Susan May, eds, *Antarctica, The Scientists' Case for a World Park* (London: Greenpeace International, 1991).
30 Kimball, L., *Southern Exposure: Deciding Antarctica's Future* (Washington, DC: World Resources Institute, 1990).
31 Kimball, *Southern Exposure*, 2.
32 Ibid., 3.
33 In the Australian newspaper *The Age* (25 April 1989), Australian Environment Minister Richardson stated a belief that mining was likely to take place, and so he wanted 'to make sure there is something in place to try, as they say in the classics, to keep the bastards honest.' Bergin, 1991, 225.
34 William Westermeyer and Christopher Joyner, *Negotiating a Minerals Regime for Antarctica* (Pittsburgh, PA: Pittsburgh University Press, 1989).
35 Ibid., 17.
36 See L. Kimball. 'Special Report on the Antarctic Minerals Convention' (Washington, DC: International Institute for Environment and Development, July and February 1988).
37 Shapley, *Seventh Continent*, 72–3.
38 Greenpeace Fact Sheet.
39 Anthony Bergin, 'The Politics of Antarctic Minerals: The Greening of White Australia', *Australian Journal of Political Science*, vol. 26 (1991): 225.
40 Ibid., 224.
41 Ibid., 225.
42 Ibid., 228.
43 Ibid., 229.
44 Biosphere reserves are an outgrowth of the Man and the Biosphere Programme established within UNESCO in 1971. Designation of Antarctica as a MAB biosphere reserve would tie the UN to the future of Antarctica to a degree undesired by the ATS membership.
45 Lynton Keith Caldwell, *International Environmental Policy* (Durham, NC: Duke University Press, 1984), 189.
46 Ibid., 107.
47 Ibid., 107.
48 In 1975, at the Eighth Meeting of Antarctic Consultative Members, New Zealand was the first country to propose making Antarctica a world park. After the minerals negotiations, in 1989, the UNGA made a similar recommendation. Shapley, *Seventh Continent*, 160.
49 Kimball, *Southern Exposure*, 15.
50 Following that meeting more states joined France and Australia in rejecting

CRAMRA, which was then considered 'no longer politically viable'. The Antarctica Project information sheets, 3 April 1991, 3.

51 The Antarctica Project information sheets 3 April 1991, 3.
52 ASOC Report on the second session of the 11th SCM, 28 August 1991 p. 9.
53 Ibid., 9.
54 ASOC report on the third session of the 11th SCM, 28 August 1991.
55 P.L. 101–594, 101–620.
56 *Sydney Morning Herald*, July 5, 1991.
57 ASOC report of the fourth session of the 11th SCM, 21 November 1991.
58 The Antarctica Project Newsletter, 1 (2), Summer 1992, 3.
59 The Antarctic Environmental Protection Act of 1992, H.R. 5459, was introduced in the US House of Representatives. On August 12, 1992, Senator Kerry introduced a similar bill, S3189, and the Bush administration forwarded its own bill, H.R. 5801, introduced by Representative Boucher. Although mark up on HR 5459 had been completed by the House Committee on Merchant Marine and Fisheries on 6 August 1992, as of late 1993, the other two pieces of legislation were still awaiting hearings.
60 Kimball, *Southern Exposure*.
61 See, for example, Lynton K. Caldwell, *Between Two Worlds: Science, the Environmental Movement and Policy Choice*, (Cambridge: Cambridge University Press, 1990); as well as Chapters 2 and 4 in this volume.
62 The Great Lakes (Chapter 4) and ivory (Chapter 5) cases illustrate how the social, especially local use, questions enter. In the Antarctic, with no resident population, the social dimension is absent.

7   Environmental NGOs in the UNCED process

Matthias Finger

In June of 1992, world leaders, journalists, and individuals proclaimed UNCED – the United Nations Conference on Environment and Development – a watershed event in setting modern society on a course of sustainable development. NGOs, many concurred, were critical to the process. What is often overlooked in the NGO role, however, is how NGOs became part of this UNCED process, as well as the fact that environmental NGOs themselves changed as a result of the process. The diversity among NGOs and the variety of strategies they followed meant that their participation in UNCED was far from uniform.

In this chapter, I argue that a fundamental tension existed throughout the UNCED process, a tension between an NGO role as defined by states and one defined by NGOs themselves. This tension was manifest in the desire of governments and of intergovernmental organizations to use NGOs as providers of data and expertise, as information disseminators, and as legitimating agents, on the one hand; and, on the other, the desire of many NGOs to use UNCED to bring about fundamental change in world development. As a result, an ongoing bargain occurred in UNCED whereby states conceded credit to some NGOs for promoting environmental and development values and, in return, gave them visibility, prominence, and sometimes even financial and logistical support. Consequently, in organizing themselves around UNCED, some NGOs acquired a certain autonomy from traditional politics and developed a new relationship with the emerging international environment and development establishment. NGO involvement in UNCED thus illustrates a process of NGO organizing, growth, and prominence observable in many contexts and international forums, some of which we describe and analyse in this book. UNCED is different in scope and degree in the sense that it considerably accelerated the process.

I begin by setting these NGO relations in the context of the ten-year history of the UNCED process. I then show how the NGO–UNCED tension played itself out in the PrepComs and the Rio meeting itself, first from the official and then from the NGO perspectives. I conclude by arguing that UNCED may have been a watershed event, but it was an event continuous with a more general trend in world environmental politics, that is, a trend away from traditional, social-movement oriented and state-centred politics. This is a trend toward a new politics, one where some NGOs play a more autonomous role *vis-à-vis* traditional political actors while simultaneously becoming 'bargaining partners' of the newly emerging international environment and development establishment.

THE BRUNDTLAND PROCESS

The historical origins of the UNCED process can be traced back to the so-called Brundtland Commission, the World Commission on Environment and Development set up in 1983. This Commission produced the 1987 Brundtland report entitled *Our Common Future*[1] and led in 1988 to the creation of the Center for Our Common Future (COCF). The Brundtland report was, in fact, the intellectual basis for the UNCED process, whereas the Center for Our Common Future became the instrument for organizing the process and including NGOs.

The Brundtland Commission itself goes back to the ten-year review conference of the United Nations Environment Programme, which, in turn, had been set up by the UN Conference on the Human Environment in Stockholm. The ten-year review was held during May 10–18, 1982, in Nairobi, and it was attended primarily by government delegates, representatives of UN agencies, and a few NGOs, in particular, the International Union for the Conservation of Nature (IUCN) and the International Council of Scientific Unions (ICSU). The delegates recommended creating a World Commission on Environment and Development. The result was the Brundtland process, the Brundtland report, and a new role for NGOs in international policy-making. Its first institutional outcome was the Center for Our Common Future.

Organization

In the autumn of 1983 the United Nations adopted Resolution 38/16, thus creating the World Commission on Environment and Development – the Brundtland Commission. The UN Secretary-General appointed Mrs Gro Harlem Brundtland of Norway, then leader of the Norwegian Labour Party, as chair and Dr Mansour Khalid, the former Minister of Foreign Affairs from Sudan, as vice-chair. The Commission was set up as an independent body that was to report back to the General Assembly in three years. It held its inaugural meeting in Geneva on October 1–3, 1984. In the mean time, Brundtland and Khalid appointed the twenty remaining members. Money to finance the Commission had to be found and a staff appointed. The canton of Geneva made office space available. In July 1984 a secretariat of fifteen professionals and clerical staff was established. James MacNeill, former Director of Environment at OECD, was appointed Secretary-General.

The members of the Brundtland Commission essentially were government officials, or scientists with government connections. Most of them were related in one way or another to the UN system and some of its specialized agencies. Among the twenty-two members of the Commission, three had United Nations Environment Programme (UNEP), two Worldwide Fund for Nature (WWF), and one International Union for the Conservation of Nature (IUCN) connections. The fact that the former Director of the United Nations Environment Programmeme (UNEP), Maurice Strong, after spending some time in private business, was also appointed as a member had far-reaching consequences, one of which was to define the role of NGOs as he and many states and international organizations saw fit. Special advisors and advisory panels were appointed. The mandate of the Commission was essentially threefold: to promote awareness, international cooperation, and action proposals. The Commission had to 'raise the level of understanding and commitment to action on the part of individuals, voluntary organizations, businesses, institutes and governments.'[2]

In preparing the report, the Commission sought oral and written input from UN agencies as well as from governments. On special occasions during fourteen site visits, deliberative meetings, and/or public information sessions, public input was sought as well. NGOs all over the world responded. The Commission and its staff processed all input and in spring 1987 published the WCED report entitled *Our Common Future* – the Brundtland report[3].

The report was released on April 27 in London. Between April and October 1987, when the report was to be adopted by the UN General Assembly, the Commission was very active. According to Warren Lindner, then Secretary of the Commission and Director of Administration:

> We worked to see to it that the report was presented to and discussed with heads of governments and senior ministers from at least one hundred countries so that the debate in the General Assembly would be an informed one. This was an effort to see that there would be adequate political support and visibility and awareness behind the report. We organized a series of regional presentations around the world not only to influence governments but also to press NGOs.[4]

The report was adopted by the UN General Assembly (UNGA) in New York the day Wall Street crashed (October 29, 1987) and thus did not get much visibility. In response, the General Assembly made a resolution, which called for follow-up conferences at national, regional, and global levels to echo the 'Call for Action' that closes *Our Common Future*. The lobbying efforts and the overall awareness raised by the report and other environmental events and activities were thus successful:

> In December 1987 a Resolution by the General Assembly called on all governments and governing bodies of organizations, bodies, and programs within the U.N. family to report in September 1989 on progress made towards achieving environmentally sound and sustainable development. Thus the

wheels were set in motion to use the Commission's report as the basis of reviewing existing programs as well as those on the drawing-boards.[5]

Officially, the WCED ceased to exist by the end of December 1987.

Unofficially, however, the Commission continued its activities in three different ways. Moreover, the core concept of the Brundtland report – sustainable development – became prominent worldwide. First, the Commission inspired the creation of UNCED and its members continued to wield influence over it: Maurice Strong, for example, who was on the Brundtland Commission, became Secretary-General of UNCED. Second, the Commission, even after its dissolution in 1987, never really disappeared and was actually reconvened shortly before the Rio conference. Finally, the Brundtland Commission transformed itself into the Center for Our Common Future.

The Brundtland report

The report provided, for the first time in such a popularized version, an assessment of the global environmental crisis. The questions of population, food production, species extinction, energy and urbanization were specifically examined. In this regard, the Brundtland Commission summarized what the different UN agencies and, during the public hearings, citizens and NGOs, had already said and written. As such, the report acknowledged and gave credence to what could no longer be disputed – the existence of a global environmental crisis.

Many observers concluded, therefore, that a major policy goal proclaimed in the report, namely, economic growth, was considerably at odds with the findings elaborated in the main body of the text. The report found that today's crisis is a crisis of human survival, especially in developing countries. Moreover, it is an ecological crisis:

> We have in the past been concerned about the impacts of economic growth upon the environment. We are now forced to concern ourselves with the impact of ecological stress, degradation of soils, water, atmosphere and forests upon our economic prospects.[6]

The solution, however, emphasizes the economic, not the human, let alone political concerns.[7]

In the report, the causes of today's global environmental problems were identified as exclusively social and political:[8] unsatisfied needs, 'in particular the needs of the world's poor' and 'limitations imposed by the state of technology and social organization on the environment's ability to meet present and future needs.' Some of these elements, identified as causes of today's global environmental problems, were also seen by the Brundtland Commission as solutions, i.e., in particular science and technology, economic growth in an open-market economy, and western-style management, especially resource management and risk management.[9]

Given this conflation of causes and solutions, and considering the nature and extent of today's global ecological crisis, it is debatable whether the proposed solutions are appropriate. In the Brundtland report the Commission acknowledged that the proposed solutions would have environmental consequences. But these consequences were systematically downplayed by considering them to be mainly acceptable risks, not long-term and often irreversible environmental degradation. Referring to the new reality where environmental degradation will threaten economic growth, the report states: 'This new reality, from which there is no escape, must be recognized and managed.'[10] Although environmental consequences of the solutions proposed were acknowledged to a certain extent, the report did not mention the social, political, and cultural consequences of these solutions, even though they are among the key topics that environment and development NGOs had brought to public attention over the past decade or so.

Although the report acknowledged that change was needed, and sometimes even suggested lifestyle changes in the North, the means that were proposed by Brundtland to bring about these changes were considered inappropriate or insufficient by many NGOs, especially southern and politically oriented NGOs, as discussed in the conclusion to this chapter. Moreover, the Commission did not describe the obstacles that have prevented such change. National governments were identified by the Brundtland report not as obstacles to change, but as key actors in this much-needed change: governments, the Commission said,[11] can deal with the sources and the effects of global environmental change, and, if they collaborate among themselves, can also manage the risks resulting from the solutions they propose.

Certain NGOs had some influence on the content of the Brundtland report, mostly via the public hearings. Importantly, NGO input is mainly visible in the analysis of global environmental problems, but not in the solutions proposed. This discrepancy is the first tangible evidence that many NGOs were welcomed for their expertise, but not for their analysis of causes and for the solutions they might derive from such analysis.

Implicitly, and sometimes more explicitly, this role of NGOs in international environmental politics had already been defined in the Brundtland report. Not surprisingly, NGOs were defined in reference to governments as comparable, in function, to the scientific community and industry. Their main role was thus to participate in development planning,[12] in decision making, and in project implementation.

Not all NGOs readily accepted this circumscribed role, however. Several NGOs, for example, did play an important role defining the concept of sustainable development. This concept appeared for the first time in 1980 in IUCN's *World Conservation Strategy: Living Resource Conservation for Sustainable Development*. IUCN viewed sustainable development as a means of natural resource conservation and management mainly at a national level; governments were to manage the resources, their uses, and the corresponding risks. Scientists, however, were to provide the necessary information and analysis to help governments make enlightened decisions. Environmental

NGOs such as IUCN were to contribute to development planning and project implementation and, more generally, to the necessary awareness-raising. The Brundtland Commission adopted this concept of sustainable development as natural resource management and transferred it from the national to the global level. According to the report, 'the principle of sustainable development . . . is at the core of Our Common Future.'[13] From the very beginning of the UNCED process, therefore, some environmental NGOs did play an important role. They provided the critical concept that became the intellectual underpinning to the Brundtland report and, from there, the UNCED process. In short, NGOs contributed to an overall social learning process (see Chapter 3).

In sum, NGOs, according to the Brundtland report, should have an input and output function. Some NGOs, such as IUCN and WWF, did participate along with science and industry representatives in the decision making. And together with industry and the scientific community, as Linda Starke wrote in her promotional book of the Brundtland report, they provided 'the services that governments are unable to'.[14] They are partners in the dialogue, as well as multiplicands of environmental awareness, carriers of planetary responsibility, and signs of hope. Or, as Lindner, director of the Center for Our Common Future, called them, 'a thousand points of light'.

Some NGOs did indeed fulfil this role. As Starke says:

> In 1990 environmental NGOs are quite likely to sit down with top government officials, especially in the industrial countries, on a regular basis. And the person across the table is now sometimes particularly familiar with NGO concerns: French Minister of the Environment Brice Lalonde was one of the founders of Les Amis de la Terre (Friends of the Earth in France). The administrator of the U.S. Environmental Protection Agency (EPA), William Reilly, was for many years head of the Conservation Foundation/World Wildlife Fund US, an influential research and funding group on environment and nature conservation issues.[15]

Before UNCED, the framework which the Center for Our Common Future sought to implement applied only to a few environmental NGOs, essentially the big American NGOs, including the Sierra Club, the National Wildlife Federation, the Natural Resources Defense Council, the Audubon Society, the Wilderness Society, the Nature Conservancy, WWF, IUCN, and, more recently, the World Resources Institute (WRI). But after UNCED this framework seems to have become more largely accepted among NGOs. In short, some NGOs, including some very influential ones, did accept the conceptual framework laid out for them by UNCED in exchange for their participation and support of the process.

The Center for Our Common Future

The activities of the Center for Our Common Future can be divided into two phases, corresponding to two different functions. From April 1988 to autumn

1989 the Center promoted the Brundtland report and the corresponding concept of sustainable development. But from the spring of 1990 onward, the Center attempted to feed the established 'Brundtland constituency' into the UNCED process. In effect, the Center was actively institutionalizing the Brundtland framework.

After the Brundtland report was submitted to the UN General Assembly in October 1987 and the Brundtland Commission officially dissolved in December 1987, the question arose as to how the Commission's work could be continued. As Warren Lindner reports:

> Ultimately it was decided that I would establish a charitable foundation called the Center for Our Common Future whose sole agenda would be to further the messages contained in the report and broaden the understanding, debate, dialogue and analysis around the concept of sustainable development. The Center would move that debate into as many sectors of society and as many countries as possible.[16]

The Center was thus established in April 1988, with voluntary funds, at the site of the previous World Commission.

The aim of the first phase of the Center was to spread the message of the Brundtland report – in particular, the idea of sustainable development. By 1992 the report had been translated into twenty-four languages and was accompanied by videos and other educational materials. The 'Basic message of the [Brundtland] report and the concept of sustainable development [was placed] on the agenda of about 600 symposia and conferences throughout the world.'[17] Finally, the Center established 160 working partners in about 70 countries. These included intergovernmental organizations and environmental development, media, youth, women's, and financial organizations, but also trade unions and professional organizations. Of course, by United Nations definition, most of these were NGOs. Among the environmental NGOs were IUCN, Worldwide Fund for Nature (WWF), the World Resources Institute (WRI), the European Environmental Bureau (EEB), and the Global Tomorrow Coalition (GTC).

> We got them to associate with the Center for Our Common Future publicly as working partners by way of making a public commitment to further the concept of sustainable development. We provide materials and resource people to the extent that we can build on the principles of the Brundtland Report.[18]

Working partners and others were called the 'Brundtland constituency' and were informed by the *Brundtland Bulletin*, which had about 3,000 subscribers.

The second phase of the Center's activities began when it became clear, in September 1989, that there was to be a UN Conference on Environment and Development in 1992. The Center redefined its mandate and priorities, to play a role in the UNCED process itself. Instead of educating the Brundtland constituency about sustainable development and publicizing the Brundtland

report, the Center's objective became the mobilization of the same constituency, including many environmental NGOs, into the UNCED process.
And when Maurice Strong was appointed Secretary-General of UNCED, the
Center shifted fully to this new role. Lindner reported: 'Maurice Strong
was appointed to head UNCED and I went to him and said we would be happy
to provide our assistance and support to mobilize in the broader constituencies.'[19] Lindner quickly transformed a previously scheduled Center
meeting in Vancouver, in March 1990:

> So we agreed to change the meeting we had planned for Vancouver into a
> first attempt to pull together groups from all over the world, from all
> different constituencies to talk about the Earth Summit in 1992 and what
> some of the basic principles of that process and its outcome ought to be.
> We had representatives of 152 organizations from 60 countries, half from
> developing countries and half from the industrialized world. Also, Mrs
> Brundtland, a representative of the Brazilian government, and Maurice
> Strong all participated in open debates with these groups on the upcoming
> 1992 Earth Summit.[20]

Maurice Strong and others must have been satisfied with the formula,
because the Center was subsequently asked to 'call a further meeting of heads
of institutions to get a mandate to play some kind of focal point role in
1992.'[21] The outcome was the so-called Nyon-meeting in June 1990, during
which the International Facilitating Committee (IFC) was created.

As it turned out, Nyon engendered a slight reassessment of the Center's
activities: from feeding the 'Brundtland constituency' into the UNCED
process, the Center now moved toward facilitating communication between
the Brundtland constituency and the UNCED Secretariat. Communication
was enhanced through the newly created *Network '92 Newsletter* and twelve
Public Forums which were held all over the world, 'where governments heard
voices of the people.'[22] The Center considered these Forums to be occasions
for citizens and NGOs to express their opinions to government delegates,
members of the UNCED Secretariat, and the PrepCom Committee. In short,
they became 'an additional avenue for input to the preparatory process for
UNCED.'[23] These Forums were in fact modelled after the public hearings the
Brundtland Commission had held to elaborate its report. But in addition to
bringing citizens and officials together in a public dialogue, Lindner made
sure that these Forums also 'raised public visibility of the UNCED process.'[24]
This is in part how the UNCED secretariat saw the role of NGOs.

As a result of the new UNCED role and the Vancouver and Nyon meetings,
the Center now helped construe NGOs as having an important input function
into the UNCED process. Furthermore, the Center actively organized them
to perform that function, that is, to feed their expertise into UNCED, but not
to challenge the premises of UNCED or the prescription coming out of the
entire Brundtland process.

THE UNCED PROCESS FROM THE OFFICIAL
PERSPECTIVE

The UNCED process officially began with UN Resolution 44/228 of December 1989. In this Resolution, the United Nations explicitly referred to the Brundtland report and decided to convene the United Nations Conference on Environment and Development (UNCED) in Brazil in 1992 to examine, among other things, 'the state of the environment and changes that have occurred since the UN Conference on the Human Environment.'[25] Decisions were made first, 'to establish a Preparatory Committee [PrepCom] open to all state members of the UN, or members of specialized agencies, with the participation of observers, in accordance with the established practice of the GA'[26]; and second, to establish an *ad hoc* secretariat. In the Resolution the secretariat was requested to draft a conference agenda, the member states were requested to prepare national reports, and UNEP, as well as other organizations and programmes of the UN system, were asked to contribute fully to the preparations of the Conference. In retrospect, one can say that UNEP never fully contributed.

The overall approach identified in Resolution 44/228 is sustainable development as laid out by the Brundtland report – that is, the creation 'of a supportive international economic climate conducive to sustained economic growth and development in all countries for the protection and sound management of the environment.'[27] Consistent with the Brundtland report, the Resolution acknowledged that science and technology will play a 'crucial role' in sustainable development, and that, therefore, 'access to environmentally sound technologies' as well as new and additional financial resources 'will be provided, especially to developing countries.' The Resolution also sought to promote 'the development or strengthening of appropriate institutions at the national, regional and global levels to deal with environmental matters in the context of socioeconomic development processes of all countries.'[28] Finally, the Resolution affirmed that 'states have the sovereign right to exploit their own resources and the responsibility to ensure that activities within their jurisdiction or control do not cause damage to the environment or other states or areas beyond the limits of national jurisdiction.'[29]

The categories of the main issues that UNCED should address were defined in the Resolution as follows:

Protection of the atmosphere (climate change, ozone depletion, acid rain)
Protection of freshwater resources
Protection of oceans and enclosed seas and rational use of living resources
Protection and management of land resources (forest, desertification, and drought)
Conservation of biological diversity
Environmentally sound management of waste (hazardous wastes, toxic chemicals, waste traffic)
Human settlements, poverty
Health

These categories, which were nearly identical to the ones used in the Stockholm process twenty years prior, became the main chapters of Agenda 21. Many NGOs, especially the more politically oriented ones, observed that certain politically sensitive but environmentally critical issues, such as population, militarization and nuclear waste, were not on the list.

Finally, Resolution 44/228 stated that the Preparatory Committee was to report to the General Assembly, [30] and not, for example, to UNEP, the 'main UN organ dealing with environmental issues.'[31] Certainly UNEP would have been much more open to NGOs than the General Assembly. The choice of UNGA highlights the different interpretation of the status and functions of NGOs between the UNGA, on the one hand, and the Brundtland Commission and its successor, the Center for Our Common Future, on the other. If the Center were more open to NGOs, especially to NGO input, the UNGA had and still has a much less friendly attitude *vis-à-vis* NGOs.

The UNCED process

On February 8, 1990, Javier Perez de Cuellar, then Secretary-General of the United Nations, announced the appointment of Maurice Strong as UNCED's Secretary-General. Under US pressure, Strong had been preferred at the last minute to Mustafa Tolba, then head of UNEP. As noted above, Strong had already been the Secretary-General of the Stockholm Conference on the Human Environment held in 1972 and was a member of the World Commission on Environment and Development. Traumatized as he had been by heavy social movement protest in Stockholm in 1972, Strong was determined from the very beginning to pre-empt any opposition to UNCED. This experience explains, at least in part, Strong's interest in NGOs.

The appointment of Maurice Strong had far-reaching consequences. With the exception of a new status and roles for NGOs, UNCED basically became, under Strong's leadership, a remake of the Stockholm process, in terms of organization and content. Indeed, the UNCED secretariat adopted the same natural resource conservation categories for the negotiations and set up the organization in much the same way he had done at Stockholm. Similar dates for PrepCom meetings and similar meeting places were chosen. Similar working structures and organizational procedures were adopted. This remake of UNCHE is evident, for example, in the projected output. UNCED, as UNCHE before it,[32] was to produce:

1 An Earth Charter, a 'series of principles to govern the relationship of people and nations with each other and with the earth.' The Earth Charter was to be the equivalent of the Stockholm Declaration. As I describe shortly, the ambitious project of devising an Earth Charter was scaled down during the negotiations, to a simple Rio Declaration.
2 An Agenda 21, 'a program of action for the implementation of the principles enunciated in the Earth Charter.' Agenda 21 was the equivalent of the Stockholm Principles of Action.

3 Financial resources, i.e., 'measures for financing the actions provided for in Agenda 21.'
4 Technology transfer, i.e., 'measures to ensure that all countries, particularly the developing countries, have access to environmentally sound technologies on an equitable and affordable basis.'
5 Institutional outcomes, i.e., 'measures for strengthening existing institutions, notably UNEP; the environmental capacities of developing agencies and organizations; the processes of collaboration and coordination among them; and the machinery to enable environment and development issues to be examined at a policy level in their relationship to other important security, economic, humanitarian and related issues.'
6 In addition and parallel to the UNCED process, negotiations were held by Intergovernmental Negotiating Committees on climate change and biodiversity.

Except for the climate change and biodiversity conventions, which have different origins and follow a different logic, the entire content and set-up of the UNCED process were thus very similar to that of the UNCHE process. The status of NGOs and the role they were scheduled to play were, however, significantly different between Rio and Stockholm.

To recall, UNCED was set up as a series of five preparatory meetings in which the Earth Charter (Rio Declaration), Agenda 21, environmental financing, and possible institutional arrangements were negotiated. As mentioned previously, climate change and biodiversity conventions were simultaneously being negotiated in two separate processes. The negotiations for UNCED took place in three separate working groups on March 5–16, 1990, in New York (preparatory meeting), on August 6–13, 1990, in Nairobi (first Preparatory Committee meeting), on March 18 – April 4, 1991, in Geneva (second Preparatory Committee meeting), on August 12 – September 4, 1991, in Geneva (third Preparatory Committee meeting), and on March 2 – April 3, 1992, in New York (fourth Preparatory Committee meeting). The Earth Summit was held in Rio de Janeiro on June 3–14, 1992. During the first ten days of the Earth Summit, negotiations were still taking place. On the last three days the Rio Declaration, Agenda 21, and financial arrangements were adopted by the heads of the member states. The institutional arrangements were handed over to the General Assembly to be decided later in the year.

All these negotiations took place exclusively among representatives of governments. NGOs, however, did have some input in the negotiation process, to which I will come shortly. Parallel to the negotiations, the UNCED Secretariat had set up ten working parties on forestry, atmosphere, environment and development, technology transfer information, biotechnology, oceans, land resources and agriculture, biodiversity, and environmental education. These working parties helped the secretariat draft the documents in between the PrepCom sessions. Among the approximately 120 members of the ten working parties,[33] nineteen were scientific experts, seventeen representatives of national government agencies, ten representatives of the United Nations Food and Agricultural Organization (FAO), nine representatives of the United Nations Environment Programmeme (UNEP), and most of the others represented other UN agencies. As for environmental NGOs, there were three representatives from IUCN, one from Greenpeace, one from the Environment and Development Action in the Third World (ENDA), one from the US Conservation Foundation (CF), and one from the US Environmental Defense Fund (EDF).

These working parties exercised a substantial influence, as they drafted the documents that were to be negotiated at the PrepCom meetings. As one can see from their sheer numbers, NGOs did exert influence, albeit limited, in the working parties as well as in the PrepCom meetings and preparations between meetings.

NGO participation from the official perspective

The General Assembly Resolution 44/228 requested 'relevant nongovernment organizations in consultative status with the Economic and Social council to contribute to the Conference, as appropriate.'[34] Moreover, the same initial Resolution invited the different states to promote a broad-based national preparatory process involving the scientific community, industry, trade unions, and concerned NGOs. To this point, this was standard United Nations practice. In a preparatory document by the Secretary-General to the UNCED organizational session in New York (March 1992), it was stated that the community of non-governmental organizations could

enrich and enhance the deliberations of the Conference and its preparatory process through its contribution/s and serve as an important channel to disseminate its results, as well as to promote the integration of environmental and development policies at the national and international levels, and that it is therefore important that nongovernmental organizations contribute effectively to the success of the Conference and its preparatory process.[35]

The secretary was, therefore, invited by the Preparatory Committee to propose arrangements for NGO participation at the first PrepCom.

The second regional conference toward UNCED – the one for Europe, i.e., the follow-up to the Report of the World Commission on Environment and Development in the ECE Region, May 8–16, 1990, in Bergen – must, in retrospect, be considered as an important step toward NGO participation in UNCED, since it was the first time that NGOs organized themselves to feed into the process. This effort signalled NGOs' intention not to simply serve as information collectors and disseminators, but to be full participants in the analysis of environmental and development problems and in the proposal of solutions.

The so-called Bergen Conference took place on a ministerial level, but there was a planned attempt to involve 'the independent sector' in the discussions with ministerial delegations. This sector included industry, trade unions, the scientific community, youth, and environmental NGOs. During the Bergen process, a new model of NGO involvement in international negotiations was actually being tested. The *Brundtland Bulletin*, which generally expressed the opinions of official UNCED, claimed that:

> The 'Bergen Process' of consensus-seeking between independent and official channels had been evolving over the two years in which Bergen was in preparation, and seems set to become the model for the 1992 process. By officially sanctioning and welcoming this kind of broad participation, Bergen represents a landmark in the democratization of the future-building process. In a paragraph of the Ministerial Declaration, the governments strongly recommend that the experiences gained in the preparation and conduct of the Bergen Conference with respect to the full involvement of non-governmental organizations, be used by other regional organizations of the United Nations and by the Preparatory Committee of the 1992 Conference.[36]

As a result of the Bergen Conference, Maurice Strong, the Secretary-General of UNCED, met with representatives 'from the independent sector, including the board of the Conference of Non-Governmental Organizations (CONGO) and other non-governmental organizations, and stressed his support for the principle of broad representation and participation.'[37] He then presented guidelines for NGO participation at the first PrepCom meeting in Nairobi and recommended that NGOs, as well as groups from a broad spectrum of society, be brought into the official process. He also urged governments, when drafting their national reports, to generate dialogues with and encourage input from all sectors of society. But NGOs suggested still other ways and means of NGO participation in UNCED, ranging from creating thirty seats for independent sector representatives at the PrepCom, to granting observer status to various NGOs at ECO '92, to placing representatives of the independent sector on all national delegations to ECO '92 as full and equal members.[38] Strong did, however, not keep his promise.

At PrepCom I in Nairobi (August 6–31, 1990) the delegates of member states debated NGO participation in the UNCED process for the first time.

> Secretary-General Maurice Strong opened the session noting that the Bergen formula for NGOs would not be 'realistic or applicable' given that the number of both the government and the nongovernment organizations would be much greater in Brazil than in Bergen. He did recommend, however, that the Bergen 'principles' be applied and suggested that NGOs could have 'their principal impact' at a national level.[39]

In other words, Strong was looking for a role for NGOs in which they would be heavily involved in providing input yet have limited impact on decision-making.

After several days of intense discussion the Preparatory Committee acknowledged that the effective contribution of nongovernmental organizations in the preparatory process was in its interest, but approved a far narrower role than the one encouraged by the Bergen Ministers.[40]

In particular, the Preparatory Committee approved the guidelines recommended to it by the Secretary-General. But it made them, in turn, subject to General Assembly approval. Indeed, the Secretary-General's recommendations went beyond Resolution 44/228, inasmuch as they also allowed NGOs not accredited to the Economic and Social Council (ECOSOC) to speak at the plenary meeting and the meetings of the Working Groups. The Secretary-General was thus asked to propose a procedure for determining NGO competence and relevance to UNCED for their accreditation. As it turned out, this procedure considerably widened NGO participation, but not necessarily NGO influence, in UNCED.

To summarize, the most relevant decisions made at PrepCom I about NGO participation can be interpreted as the following:[41]

1 Non-government organizations were to have no negotiating role in the work of the Preparatory Committee.
2 Relevant NGOs might, at their own expense, make written presentations in the preparatory process.
3 Relevant NGOs in consultative status with the Economic and Social Council might be given an opportunity to briefly address plenary meetings of the Preparatory Committee and meetings of the Working Groups. Other relevant NGOs might also ask to speak briefly in such meetings. Any oral intervention by an NGO would, in accordance with normal United Nations practice, be at the discretion of the chair and would require the consent of the Preparatory Committee or the working Group.

The General Assembly of the United Nations at its forty-fifth session endorsed the *ad hoc* rules of the Nairobi PrepCom I meeting, agreeing, therefore, to PrepCom accreditation of NGOs not in consultative status with ECOSOC. At the second PrepCom meeting the Secretary-General still presented guidelines for 'determining NGOs' competence and relevance to the work of PrepCom', which were accepted by the Preparatory Committee. As a result, accreditation was offered to virtually every NGO that applied; only four applications were denied.

Overall, it appears that NGO accreditation to the UNCED process was different from United Nations standard practice only inasmuch as NGOs in non-consultative status with ECOSOC were granted the same rights as the ones with consultative status. It turned out that NGOs with non-consultative status in ECOSOC were indeed the majority of the NGOs involved in UNCED. As a result of this quite easy access, NGO participation in UNCED was at a record high, compared to other UN negotiations and conferences. Therefore, UNCED had an unprecedented number of participating NGOs in the UN system. Record numbers are not equivalent to impact, however. Large numbers served the organizers' purposes well but may have actually hindered the NGO community, especially that segment of the community which tried to address underlying causes and to propose meaningful solutions.

NGOs were accredited several times starting with PrepCom II. At the end of PrepCom IV the total number of accredited NGOs to the UNCED Secretariat was 1,420, in addition to NGOs already accredited to ECOSOC. Such accredited NGOs represented a wide range of sectors, namely professional organizations, trade unions, women and youth organizations, environment and development NGOs, religious organizations, business and industry, scientific organizations, and many others. Notably underrepresented, however, were the political greens composed of environmental movement organizations and green parties in Europe.

NGO participation during the various PrepCom meetings increased throughout the UNCED process. Nevertheless, recalling that 'NGOs shall not have any negotiating role in the work of the Preparatory Committee',[42] most of the accredited NGOs did not actually participate in the negotiations. Generally, NGO participation remained limited to formal sessions, that is, sessions in which government delegates, representatives of international organizations, and NGOs made statements for the record. The secretariat's policy about NGO access to formal-informal and informal-informal meetings was not consistent.[43] Informal meetings, sometimes referred to as formal-informals, were interpreted but not transcribed. In these informal meetings, NGO access was determined by the chair. NGO observers were not allowed to make statements. Informal-informal meetings were conducted in English only and ranged from open-ended meetings held in conference rooms sometimes open to NGOs (depending on the Working Group chair's decision) to small meetings held in the chair's office, involving only a limited number of delegates. In all, questions of accreditation, access, and participation of NGOs in the UNCED process were only the concrete manifestations of attitudes and intentions held by Strong and the UNCED secretariat about the status and role of NGOs.

The secretariat's idea of NGOs

Although the secretariat, in accordance with UN practice, did not use the term 'independent sector', its use of the term NGO was more or less synonymous with that used by the Brundtland report and the Center for Our Common Future.

> It should be remembered that in the guidelines adopted, the Secretary General pointed out that the term 'nongovernment organizations' includes groups from industry, science, trade, environment and development, youth and women, as well as those NGOs with and those without consultative status with the Economic and Social Council.[44]

The implicit assumption behind the term 'independent sector' was that the member state or, in this case, the UNCED secretariat, was considered to reflect the *public* interest, whereas the various independent sectors or NGOs expressed and aggregated *private* interests. In the view of the UNCED secretariat, as in the view of the Brundtland Commission, NGOs should have an input and an output function for the official government process. As stated in an UNCED secretariat's promotional brochure:

> [each] NGO could enrich and enhance the deliberations of the Conference and its preparatory process, and could serve as an important channel to disseminate its results, as well as to promote the integration of environment and development policies at the national levels.[45]

During the UNCED process the secretariat had stressed the input function of NGOs. Consequently, the UNCED secretariat was looking for so-called working partners, that is, coalitions of the various independent sectors. To feed these partners into the UNCED process, the UNCED secretariat asked various sectors to organize themselves to speak in one voice. This model of channelling NGOs into the UNCED process in fact worked well for some independent sectors, such as science, women, and indigenous people, but less well for others. Development and, especially, political environmental NGOs were much less comfortable with this model and, consequently, participated much less in the UNCED process. Their main critique was that they were only allowed to contribute information, but not to participate meaningfully in the decision-making.

By contrast, this model of NGO participation suited the business and industry sector well. In fact, this sector seems to have heard the Brundtland Commission's call for sustainable development before all other independent sectors. Consequently, it became the best organized sector, the one that fed most directly and most efficiently into the UNCED process. The Brundtland Commission had scarcely begun when in 1984, the International Chamber of Commerce (ICC), in collaboration with UNEP, organized the first World Industry Conference on Environmental Management (WICEM I) in Versailles. As a result, the International Environmental Bureau (IEB) emerged in 1986. IEB was located first with the World Economic Forum in Geneva and is now with the ICC office on Environment and Energy in Norway. IEB is a trans-industry clearing-house on environmental management information.

The Bergen Conference was the next significant step in the business sector's endeavours. Out of the Bergen Conference and the parallel Industry Forum came the European Green Table, a 'contribution to the work of ICC towards the 1992 UNCED.'[46] ICC was also mandated in Bergen to prepare several industry projects. These formed the core of an industry initiative that was finalized at WICEM II. This initiative prepared 'the main policy issues relevant to world industry in relation to UNCED.'[47]

The participation of the business sector was enhanced when, in 1991, Maurice Strong appointed Dr Stephan Schmidheiny, a Swiss industrialist and billionaire, as his principal business advisor for UNCED.[48] Schmidheiny recruited a group of forty-eight business leaders from around the world and created the Business Council for Sustainable Development (BCSD) during the second World Industry Conference (WICEM II) held in Rotterdam in April 1991. The council provided advice and guidance to the UNCED secretariat on initiatives and activities undertaken by business and industry with respect to the preparatory process for the 1992 Conference, including programmes developed by the International Chamber of Commerce and other business organizations and bodies, programmes developed by the World Economic Forum and its Industry Fora, and programmes developed by individual corporations and business leaders.[49] Moreover,

> Mr. Strong requested that the mandate be carried out well in advance of the Earth Summit so that the input to the Business Council's members could be taken into consideration during the consultative process that the UNCED Secretary General is carrying out prior to Rio.[50]

In other words, the BCSD fed directly into the consultative process of UNCED, whereas most NGOs fed, if at all, into the discussions at the Preparatory Committee's meetings. Clearly, the difference in venues and access was significant because the implicit model used for NGO access was a lobbying model which favoured financially potent and organizationally strong lobbyists. Although access had indeed been made easy for all NGOs – including environmental NGOs who spoke with one voice and did not challenge the conclusions of the Brundtland report – business and industry, along with the big northern conservationist environmental NGOs, actually got the best access to the power-brokers of UNCED.

UNCED FROM THE PERSPECTIVE OF ENVIRONMENTAL NGOs

Let me now turn to the UNCED process as seen from the perspective of NGOs themselves, especially environmental NGOs. What roles did they play in UNCED? How did they try to, or effectively come to, influence the UNCED process and its outcomes? Above, I have shown that the Center for Our Common Future and the UNCED Secretariat saw NGOs mainly as potential constituencies, as groups who, given the premises and conclusions of Brundtland, should organize themselves to provide input into the UNCED process. I have also shown that this worked particularly well for business and industry. But how did it work for environmental NGOs? In this section I examine how environmental NGOs mobilized and organized for UNCED and assess how and whether they actually influenced both the UNCED negotiations and the evolution of the global NGO community.

NGO coalitions mobilizing for UNCED

UNCED forced environmental and other NGOs to mobilize their con-stituencies and organize themselves in order to have a chance to make an input into the process. The creation of NGO coalitions to mobilize or facilitate access to UNCED was, indeed, the most striking phenomenon in the environmental NGO community. In its scope and nature, this coalition-building was unprecedented. One can identify, around UNCED, two such efforts on a global level. The first effort was the Center for Our Common Future's attempt to facilitate NGO access to the UNCED. For this purpose the Center created in 1990 the International Facilitating Committee (IFC). The second effort was a reaction from some NGOs against the IFC's efforts, i.e., the attempt by the Environmental Liaison Committee International (ELCI) in Nairobi to build a parallel social movement, or citizens' summit. Besides these two efforts, there were numerous attempts to mobilize NGO coalitions on national and regional levels.

The International Facilitating Committee

The IFC was an offshoot of the Center for Our Common Future. To recall, the Center's original role was to popularize the Brundtland report. But after the UNCED process was announced, the Center shifted its focus to feed the Brundtland constituency into the UNCED process. In March 1990 the Center convened in Vancouver representatives of 115 independent sector organizations from more than forty countries to consider their respective involvement in UNCED.

Parallel to the Center's efforts, the European Environmental Bureau (EEB), a coalition of lobbying environmental NGOs at the EC level, organized a preparatory NGO meeting to feed into the European regional Conference on Environment and Development in Bergen in May 1990. Of the 350 NGOs invited on a boat trip between Vienna and Budapest (March 19–21, 1990), fifty NGO representatives were selected to go to Bergen. There they elaborated a common Agenda for Action. As noted at the ECE meeting in Bergen, NGOs obtained observer status and the right to contribute to the Bergen ministerial meeting. Yet NGOs were excluded from the ministerial session that followed the working session. One-third of the thirty-four governments present at Bergen, however, included NGO representatives in their official delegations.[51] In their final report, the ministers recommended that

> the experience gained in the preparation and the conduct of the Bergen Conference with respect to the full involvement of nongovernmental organizations be used by other Regional Commissions of the United Nations and by the Preparatory Committee of the 1992 Conference on Environment and Development.[52]

As discussed above, the Bergen meeting was considered by the UNCED Secretariat and the Center for Our Common Future to be a model for NGO participation in the entire UNCED process.

Therefore, in June 1990 the Center for Our Common Future convened a strategy meeting for the independent sector, including environment and development NGOs, business and industry, trade unions, professional associations, scientific and academic institutions, women's organizations, youth groups, religious and spiritual groups, indigenous people's organizations, and other citizen groups in Nyon, Switzerland. Their task was to prepare for NGO participation in UNCED. Participants at that meeting decided to create a new body to coordinate NGO activities for UNCED, stipulating that it would serve a facilitative function rather than a representative function for the global NGO community. It would be structured according to constituencies or independent sectors. NGO participants debated whether industry should be part of the independent sector and finally accepted it. The IFC, thus, was created as a coalition of independent sectors. It was to be located at the Center for Our Common Future in Geneva, but remained financially independent from it. In collaboration with the Center for Our Common Future, the IFC produced the *Network '92* newsletter.

As a result, the IFC was a coalition or, maybe better, a patchwork, of various independent sectors, themselves represented by particular organizations, such as IUCN, EEB, ICC, CNN, ICSU, the Asian NGO Coalition (ANGOC), and so forth. Consequently, it was very difficult for the IFC to agree on anything substantive except, perhaps, the call for sustainable development and the active participation of independent sectors in the UNCED process. It is, therefore, not surprising that some NGOs – in particular, the more social movement oriented ones – were unhappy with the IFC and created their own alternative.

In its mission statement, IFC stated that it sought to assist organizations and networks in the independent sector to define their roles in UNCED, to promote fair and effective participation in UNCED on behalf of the independent sectors, and to provide a forum for dialogue among the independent sectors. The IFC physically facilitated NGO access to UNCED, organized information briefings for NGOs before and after PrepCom meetings, and held the parallel '92 Global Forum at UNCED in Rio, which was, in effect, the culmination of the IFC's efforts. The Global Forum was attended by approximately 20,000 people; 1,600 organizations actively participated. Overall, however, the Global Forum was a show, rather than an input into the UNCED process. It was a forum parallel to and separate from the UNCED Conference where NGOs and more commercially oriented organizations could exhibit their concerns and actions, get together, and plan future actions. Its direct input into the UNCED process was minimal. The one exception, to which I turn shortly, was the International NGO Forum in which a significant segment of the NGO community attempted to set an alternative agenda of action.

The Environmental Liaison Committee International

All along, the IFC was unable to dissipate concerns of some of the more social-movement oriented NGOs, especially ELCI and Southern grass-roots organizations. These NGOs criticized in particular the role business and industry had come to play in the IFC and in UNCED in general. For example, at the Nyon Meeting of the Center for Our Common Future (June 1990), at which the IFC was created, the IFC decided to organize, among other things, a pre-UNCED meeting of all interested groups, to be held approximately six months before the Conference to concentrate on the official Conference agenda. At that time, however, the ELCI 'board members present felt that the ELCI, and not the IFC, should organize such a pre-Brazil meeting and that the IFC should only play a facilitating role.'[53] ELCI's opposition is understandable, given its prominent role over many years in dealing with issues central to environment and development. In the end, ELCI quit the IFC and began to define its own strategy for the Rio Conference.

An International Steering Committee, set up to guide the ELCI's work, met for the first time on August 4, 1990, and decided that 'ELCI will sponsor national and regional consultations on environment and development, the first one being scheduled for early December 1990 in India, and will gather the reports of these consultations into a Brazil document.'[54] This International Steering Committee was co-chaired by the Brazilian NGO Forum (BNGOF) and Friends of the Earth (FoE). Of the sixteen members of the committee, six were from ELCI, one from IUCN, and one from ENDA. The others were not direct representatives of environmental NGOs. All represented NGO coalitions. The main focus of the International Steering Committee was to identify local solutions to global problems that can contribute, in particular, to changes in lifestyle and consumption patterns. The ideological orientation of this effort by ELCI, FoE, BNGOF was radical compared to that of the IFC. The focus was on grass-roots and people-oriented initiatives, much of which would be in opposition to governments. The approach was, therefore, also much more confrontational.

In November 1990, the International Steering Committee accepted the French government's offer to sponsor a global NGO conference, to be held in December 1991 in Paris. The French offer was conditional upon the Steering Committee's selection of 850 participants among environment and development NGOs. The so-called Paris NGO Conference took place during December 17–20, 1991. This Conference produced an NGO position paper entitled 'Roots for Our Future', which contained, among other things, a synthesis of NGO positions and plan of action that dealt, in particular, with climate change, biodiversity, forestry, biotechnology, the General Agreement on Trade and Tariffs (GATT), resource transfer, institutions, and lifestyle. It focused on what grass-roots organizations can and should do. This document was to be the social-movement sector's input into the Rio Conference. It was to lead to an NGO/social-movement gathering to be held in Rio parallel to the official Conference and even parallel to the 1992 Global Forum itself. Early on, this gathering was called the International Civil Society Conference; its final name was the International NGO Forum.

This International NGO Forum took place in Rio within the organizational framework of the Global Forum organized by the IFC; of all the NGO events, this one had the most substance. Many NGOs, especially the politically oriented environment and development NGOs, participated. They elaborated position papers, statements, and alternative treaties. The alternative treaty writing process was considered successful by those who participated, although no immediate impact upon the UNCED outcomes can be detected. However, since Rio, many groups who wrote or signed on to these treaties have continued the process initiated in Rio by organizing meetings to refine and develop their action plan in accordance with the treaties. Some of the NGOs are using the treaties in their advocacy and public mobilization work.[55]

Other efforts to feed NGOs and citizens into the UNCED process

IFC's and ELCI's endeavours to mobilize NGOs and feed them into the UNCED process were certainly the most prominent. One must, however, mention similar efforts on regional levels, the most important of which were the Global Education Associates' Earth Covenant Project, a people's treaty on environmental conduct, the Global Tomorrow Coalition's Globescope Americas Assembly, a conference on sustainable development held in Miami from October 29 to November 2 1991, the World Resources Institute's New World Dialogue on environment and development in the Western Hemisphere, a dialogue among citizens and leaders of North and South America to draw up a North–South compact calling for specific commitments by governments to reduce greenhouse gas emissions, halt forest and biodiversity loss, slow population growth, increase food production, and tackle problems of urban and industrial pollution.

On a national level there were the efforts of the Brazilian NGO Forum, a coalition of Brazilian NGOs, the US Citizens' Network on UNCED, the Canadian Participatory Committee for UNCED, and the Norwegian Campaign for Environment and Development, among the most visible national NGO coalition-building efforts. But many other NGO coalitions were created especially to participate in the UNCED process on a national level. Still other NGOs reoriented their activities to accommodate the UNCED process.

All these national, regional, and international efforts to mobilize and organize NGOs led them to create their own niche in the UNCED process. But rarely did the outcomes of these NGO coalitions at national, regional, or global levels feed directly into the ongoing negotiations at UNCED. With the exception of the International NGO Forum, most of these NGO coalition efforts took the form of appeals to national governments or world leaders. They asked that their statements be read or distributed in the official meetings in the preparatory process, or in Rio. In general, these NGO coalitions sought to raise awareness, promote sustainable development, and ask for further NGO participation. Overall, the mobilizing efforts certainly increased the visibility of the UNCED process, as well as their own visibility.

Over time, it also became clear that the public relations aspect of UNCED, to which NGOs mostly contributed, became more and more separated from its negotiation aspect. Most of the NGO coalitions created for and around UNCED were not actively participating in the negotiations of the UNCED process. They seemed to be satisfied with the overall awareness-raising and promotional role they had come to play. As a result, only a few environmental NGOs actively sought participation in UNCED negotiations.

Influencing the negotiations

Throughout the UNCED process, many environmental NGOs actively tried to influence negotiation outcomes. They sought to influence the wording of the Rio Declaration on Environment and Development, of Agenda 21, of the planned agreement on financial mechanisms, and of the agreement on ways to reform and strengthen international institutions. NGOs could influence the UNCED negotiations in two ways: by lobbying during and between PrepCom meetings, and by participating on national delegations.

Lobbying negotiators

Lobbying at and between PrepCom meetings included making statements in the plenary meeting and/or in the three Working Groups, drafting concrete proposals, influencing government delegates, briefing government delegates at the PrepCom meetings, and becoming members of the working parties. In principle, negotiations were not open to NGOs; therefore, a lot of lobbying went on in the hallways and in the coffee lounge. NGOs' effectiveness depended considerably on the resources they could marshall and on how well statements or alternative draft proposals were drafted and timed so that they could be taken up by the negotiators.

This lobbying model thus favoured those NGOs with funds and experienced lobbyists – namely, the big NGOs of the North. These NGOs often had concrete proposals related to their respective fields of specialization and were, consequently, most prominent. On the topic of biodiversity, for example, environmental NGO lobbyists such as WRI, IUCN, WWF, and the Coalition to Protect the Earth (CAPE '92), an NGO coalition of the six largest US environmental organizations, were most visible. To a certain extent, indigenous and Third World groups, especially when it came to questions of intellectual property rights, also participated. On climate change, one could find NGOs such as WRI, the Climate Institute, the Climate Action Network (CAN) created by EDF, Greenpeace, CAPE '92, and others. Depending on the issue, NGOs did specialize but the large, well-financed, mostly Northern groups were the most prominent.

It is worth noting that almost all lobbying environmental NGOs were active within the framework of the various topical areas defined by UNCED. Very few lobbied on the crucial cross-cutting issues of technology transfer, environmental financing, and institutional reform, although NGOs did lobby in the cross-cutting fields of poverty (mainly Third World and development NGOs), population, women, and indigenous people. It is difficult to assess the exact impact of the various environmental lobbying groups' activities on the negotiation documents. But analysis of Agenda 21 and of the Rio Declaration shows that NGOs influenced the wording only marginally. If NGO wording was adopted at all, it was incorporated into the texts in an almost token manner and in disregard of NGOs' underlying arguments.[56]

NGO representatives on government delegations

NGOs' second strategy to influence the direct outcome of the UNCED negotiations was to have individual members serve on their respective national delegations. From one PrepCom to the next, and eventually at Rio, an increasing number of countries actually appointed representatives of the independent sector to their delegations. They represented business and industry and research institutes, but sometimes also environment and development NGOs.

Dawkins reports that:

Canada was the first country to put NGO representatives on its national delegation. This occurred during PrepCom I in Nairobi, where Canada was apparently the only country doing this. Moreover, Canada set another precedent by letting the NGO representative speak in Plenary. By PrepCom II at least eight countries had appointed NGO representatives to their delegations, almost all from the North. They were Australia, Canada, Norway, the Netherlands, the United Kingdom, the United States, the Union of Soviet Socialist Republics, and India.[57]

The number of counties including environmental NGOs on their delegations increased somewhat during the UNCED process, and was probably highest in Rio. These countries included Norway, Sweden, the United Kingdom, Denmark, Finland, Canada, New Zealand, the United States, Australia, the Netherlands, the Commonwealth of Independent States (CIS), India, Switzerland and France.

Moreover, besides sitting on government delegations, influencing national reports, and lobbying negotiators, environmental and other NGOs were also active in many other ways during and between the PrepCom meetings. Representatives attended the PrepCom meetings, sat on NGO task groups dealing with specific topics, disseminated and shared information via NGO publications such as *Cross Currents*, and participated in computer networking.

In this section we have seen how most environment and development NGOs have conformed to UNCED's lobbying model: they have mobilized and organized. At times, they have had an impact on the UNCED documents, albeit a limited one. But their main contribution certainly was to develop what we call in this book 'NGO relations'.

UNCED AND THE EMERGENCE OF INTERNATIONAL
ENVIRONMENTAL NGOs

Overall, the official outcome of the two-year UNCED process can scarcely be called a success. In fact, the process essentially failed to meet its own objectives in all the main categories of negotiations and their outcomes.[58] The overall thrust of the UNCED documents became the promotion of economic development, rather than environmental protection. The Rio declaration contains generalities, while Agenda 21 is an 800-page document, saying everything and thus nothing; it is almost impossible to implement. Substantial topics like the environmentally destructive impacts of economic growth, technological fixes, and militarization, among others, were left out. Finally, of the $60 billion that were considered necessary by the UNCED secretariat to achieve sustainable development, only $2 billion were raised.

Environmental NGOs, however, can hardly be made responsible for this failure. What is relevant to NGO relations, is not so much NGO influence on UNCED's official negotiated outcomes, but UNCED's impact on the NGO phenomenon. UNCED substantially accelerated the process of international environmental NGO growth, a process documented elsewhere in this book. UNCED forced environmental and other NGOs to organize. What is more, to be credible before their own constituencies, they had to strategize and collaborate with each other, if for no other reason than that they would be left behind without such alliances. As such, UNCED helped strengthen the already existing international environmental NGOs. It helped create linkages among NGOs that previously had been mainly operational only at the national level.

UNCED thus established new and different kinds of NGO relations, significantly contributing to the transcendence of many NGOs from traditional politics, as well as from the environmental movement from which they originated. They did this, moreover, by entering bargains with emergent international environment and development actors. Southern NGOs, for example, used UNCED to gain visibility, prominence, and sometimes even free trips and other support, as well as access to power. Through UNCED these NGOs established relationships with the most vocal governments in UNCED, in particular the United States, Canada, India, Pakistan, and Malaysia, with UN agencies, and even with industry, especially transnational corporations. Overall, NGOs as a category of actors in international environment and development politics have improved their status and their bargaining power. By asserting themselves as international environmental actors, these NGOs have gone beyond traditional national politics, as well as the environmental movement from which they originated.

At the same time, although many gained access to the highest levels of international environmental politics, the environmental movement after UNCED is more fragmented than before, splitting into at least three distinguishable factions. The first faction includes the mainstream environmental NGOs, such as WWF, IUCN, the World Resources Institute (WRI), and the 'Big Six' in the US, most of which have operated internationally for some time. As discussed, these are the ones which struck mutually agreeable bargains with major actors in UNCED. As UNCED unfolded, they increased their access to the secretariat and to the delegates of Northern countries. They were often consulted and sometimes even had representatives at crucial positions within the UNCED process. They are the prototype of the international environmental organizations which have gone beyond traditional state-centred politics and the environmental movement. Although their credibility grew in the eyes of many, they have, at the same time, become somewhat co-opted and have isolated themselves from the rest of the environmental movement. As a result of UNCED, they seem to have increased their power internationally. But by working closely with governments and international development agencies, they probably also have become more like them.

The second faction of NGOs emerging from UNCED include the political environmental NGOs, represented most prominently by ELCI, Friends of the Earth, the European Environmental Bureau, and the Brazilian NGO Forum. Throughout UNCED they deliberately kept close ties with the environmental movement. They moreover tried to establish links between the environmental movement in the North and environment-oriented NGOs in the South. Their relationships to the governments and the international environment and development agencies is much more ambiguous than that of the first faction. At times, these politically oriented NGOs entered into dialogue and negotiations with actors of the official UNCED process, but at other times they were also confrontational. For example, ELCI had taken up the French government's offer to pay for a pre-UNCED meeting in December 1991. Yet, after the meeting there was no follow-up in government–NGO collaboration. The Brazilian NGO Forum had an equally ambiguous relationship with the Brazilian government, and the same is true for the relationship between the European Environmental Bureau and the EC. Much of this ambiguity is due to the fact that these political NGOs are all engaged politically in their respective countries, often in a social movement capacity. Also, because of their primary commitment on a national level, as opposed to an international level, these NGOs displayed significant ideological and political differences among themselves, mostly over political strategy but sometimes also over the question of environment versus development.

As a result of all these differences, the politically oriented NGOs did not manage to organize themselves as stable players and working partners of the environment and development establishment in the same way as the first faction of NGOs. Also, being more politically oriented, there was quite naturally conflict over leadership among these NGOs, in particular between environmental NGOs from the North and development-oriented NGOs from the South. By attempting to force all NGOs to organize and speak in one voice, UNCED exacerbated differences and other latent conflicts among these political NGOs. Nevertheless, as one could see from the international NGO Forum held in Rio, these political NGOs have most significantly contributed to the international environmental NGO phenomenon we have been high-lighting throughout this book.

Two international environmental NGOs – the Third World Network and Greenpeace International – must especially be mentioned here. Although similar to the first faction of international environmental NGOs with respect to their organizational structures and effectiveness, ideologically they clearly belong to the second faction. The Third World Network and Greenpeace International lobby traditional governments and, at the same time, transcend narrow, traditional politics. They are, consequently, among the most typical representatives of the newly emerging international environmental NGO phenomenon.

Finally, it was possible to observe a third faction of NGOs at UNCED which, from a purely quantitative point of view, may well be the most important faction. Unlike the NGOs of the first faction, these NGOs were not organized at an international level. And unlike the NGOs of the second faction, they had no political agenda, much less did they seek to transcend traditional politics. This is the type of NGOs whose main focus is con-sciousness-raising on specific issues ranging from sea turtle protection to the necessity of planetary spiritual healing. They saw in UNCED an opportunity to raise consciousness on a much larger scale than previously imagined. In other words, UNCED for them was a forum to speak to the world. But since these numerous NGOs did not lobby, nor mobilize or build coalitions, they were not particularly effective. As a result, very few of them established long-lasting transnational relationships with other NGOs. Thus only rarely did some of these NGOs illustrate the type of phenomenon we have characterized in this book as emerging NGO relations.

CONCLUSION

In this chapter I have analysed environmental NGOs' involvement in the UNCED process. I have shown how UNCED has considerably accelerated and highlighted the very process this book is all about, namely the process of growing environmental NGO prominence and strengthened NGO relations. We have seen how NGOs have been actively disseminating environmental awareness and how, by doing that, many of them have become agents of social learning. But the main focus of this chapter has been to show how the very way NGO participation was conceived and set up in UNCED led some environmental NGOs to further transcend traditional politics and become independent bargaining partners with governments and international agencies. This is particularly true, as I have tried to show, of the politically oriented NGOs and to a lesser extent also of the traditional big international environ-mental NGOs.

In short, UNCED has significantly contributed to promoting a new type of actor in international environmental politics – that is, a type of actor that links local environment and development concerns with global considerations on the one hand, and that establishes a relationship between biophysical concerns and political considerations on the other.

NOTES

1 World Commission on Environment and Development, *Our Common Future* (Oxford: Oxford University Press, 1987).
2 Ibid., 357.
3 Ibid.
4 Warren Lindner, cited in Stephen Lerner, ed., *Earth Summit. Conversations with Architects of an Ecologically Sustainable Future* (Bolinas, CA: Common Knowledge Press, 1992), 237.
5 Linda Starke, *Signs of Hope. Working towards Our Common Future* (Oxford: Oxford University Press, 1990), 7.
6 *Our Common Future*, 5.
7 *Signs of Hope*, 63.
8 Ibid., 43.
9 For an elaboration of these points, see Chatterjee, P. and M. Finger, *The Earth Brokers: Power, Politics, and World Development* (London: Routledge, forthcoming), especially chapter 2.
10 *Our Common Future*, 1.
11 Ibid., 21, 313–42.
12 Ibid., 21.
13 Ibid., 8.
14 *Signs of Hope*, 65.
15 Ibid., 66–7.
16 Lindner, cited in *Earth Summit*, 241.
17 *Idem.*
18 *Idem.*
19 *Idem.* 242.
20 *Idem.*
21 *Idem.* 243.
22 *Brundtland Bulletin*, October 1990, 9.
23 *Brundtland Bulletin*, September/December 1990, 7.
24 Lindner, cited in *Earth Summit*, 245.
25 UN General Assembly Resolution 44/228, 5.
26 Ibid., 8.
27 Ibid., 4.
28 Ibid., 7.
29 Ibid., 4.
30 Ibid., 10.
31 Ibid., 9.
32 See UN Document A.C. 151/PC, 14.
33 *Network '92*, July 1991, 6–7.
34 UN General Assembly Resolution 44/228, 9.
35 UN Document A.C. 151/PC, 2.
36 *Brundtland Bulletin*, June 1990, no.8, 1.
37 *Network '92*, August 1990, no.1, 2.
38 *Network '92*, August 1990, no.1, 2. 'Eco '92' is an early acronym for the 'Earth Summit'.
39 Kristin Dawkins, 'Sharing Rights and Responsibilities for the Environment. Assessing Potential Roles for Non-Governmental Organizations in International Decision-Making' (masters thesis, MIT, Cambridge, MA, 1991), 45.

40 Ibid., 46.
41 UN Document A. 45/46, 22.
42 Ibid.
43 Johannah Bernstein, *et al.* 'PrepCom III: Third Week Synopsis', NGO strategy group paper, Geneva, 1991, 3.
44 *Network '92*, no. 2, 1.
45 UNCED Secretariat promotional brochure, 1991, 2.
46 *Brundtland Bulletin*, September/December 1990, SF18.
47 *Brundtland Bulletin*, March 1991, 69.
48 *Network '92*, October 1990, 1.
49 *Brundtland Bulletin*, October 1990, 11.
50 *Network '92*, April 1991, 2.
51 Dawkins, 'Sharing Rights and Responsibilities', 45.
52 UN Document A. Conf. 151/PC, 10.
53 Dawkins, 'Sharing Rights and Responsibilities', 42.
54 *Brundtland Bulletin*, September 1990, 53.
55 *The Independent Sectors Network*, No. 29, August 1993 (6).
56 Chatterjee, P. and M. Finger, *The Earth Brokers: Power, Politics, and World Development* (London: Routledge, forthcoming), especially chapters 4 and 5.
57 Dawkins, 'Sharing Rights and Responsibilities', 48.
58 For a critical analysis of the UNCED outcomes, see: Finger, M. 'Politics of the UNCED Process', in Sachs, W., ed., *Global Ecology. A New Arena of Political Conflict* (London: Zed Books, 1993) 36–48; Chatterjee, P. and M. Finger, *The Earth Brokers: Power, Politics and World Development* (London: Routledge, forthcoming).

Part III
Conclusion

8   Translational linkages

Thomas Princen, Matthias Finger and Jack P. Manno

In this book we have examined the NGO phenomenon from two theoretical perspectives and we have described and analysed four case histories of world environmental politics involving international environmental non-governmental organizations (NGOs). As stated in Chapter 1 and elaborated in Chapters 3 and 7, a premise of this book has been the persistence and growth of an increasingly global ecological crisis. The NGO phenomenon arises at a time when existing state institutions are reactive at best to the crisis, and economic globalization appears to contribute significantly to its acceleration. Because the NGO phenomenon arises coincidentally with these trends – global environmental crisis and economic globalization – it is tempting to explain the NGO phenomenon as a logical response: the crisis requires more and stronger environmental advocates, and consequently NGOs have proliferated and grown. Such a functional argument too easily ignores the peculiarities of NGO relations at the international level and, in particular, the political implications of the biophysical conditions of the crisis.

Consequently, we have attempted to explain the NGO phenomenon by demonstrating how NGOs perform key roles as independent bargainers (Chapter 2) and as agents of social learning (Chapter 3). With the addition of the case studies we can now discern some of the general characteristics of the ecological crisis that creates the political space, the niche, in which NGOs perform these roles. From these roles and characteristics, then, we can construct a set of propositions that characterize NGOs' distinctive contribution to world environmental politics.

In this chapter, therefore, we outline the general characteristics of ecological crisis and their political implications and then develop a theoretical construct that connects many of these characteristics. We argue that the distinctive contribution NGOs make is to draw out the political implications of biophysical trends at the local and global levels and to challenge the limitations of the traditional state-centric system. Finally, we suggest applications and directions for future research.

THE NGO NICHE: THE ENVIRONMENTAL CONTEXT

Each of the four cases treated in this book illustrates one or more aspects of the growing global ecological crisis and the implications for traditional politics and the role of international environmental NGOs. In particular, these cases and much contemporary literature suggest that, with the acceleration of the global ecological crisis, transboundary physical and biological processes can no longer be ignored.[1] Small, dispersed ecological disturbances tend to combine and interact, often in subtle ways. Many result in non-linear, synergistic and threshold effects, with episodic, unpredictable, and system-wide consequences. For example, toxics resulting from widespread industrial and consumption patterns can bioaccumulate in fish, threatening ecosystems and human health. Small-scale tourism can wreak irreversible damage in an ecosystem like Antarctica where the foodchain has only a few links. Overharvesting can be only marginally apparent until dramatic population declines ensue. Overall, large-scale processes can build momentum in such a way that effects are not observable until they are irreversible. What is more, an increasing number of local, regional and even national ecological problems are traceable to non-local causes. Income growth in Japan drives poaching in Africa, or global warming threatens to release submarine methane in the Antarctic ecosystem.

Many of these problems transcend national boundaries either because they cross boundaries or because their causes and effects occur on all sides of borders. Pollutants in the Great Lakes, habitat changes in Africa, and ozone depletion over Antarctica are illustrative. In such cases, both the sources and the damage belong to two or more states. And, most important for understanding the social effects, on all sides of the borders some actors benefit from the economic activity which is the source of the damage and some see themselves as aggrieved. Hence, both the causes and the effects are transnational.

The biophysical characteristics of transnational environmental problems and their social impacts call for a new kind of politics. Chapters 2 and 3 highlighted the inadequacies of traditional politics. In light of the case studies, we can now expand our understanding of these inadequacies by explicitly grounding them in the biophysical conditions. To elaborate the distinction between traditional and new environmental politics, we characterize traditional politics as domestic mediation and diplomatic compromise. We then show how they are fundamentally different from the newly emerging politics, namely, that associated with global environmental crisis.

As argued in Chapter 3, traditional politics is a nationally oriented process where states mediate among conflicting interests. Domestically, the classic environmental conflict between upstream polluters and downstream recipients is mediated by the state through the allocation of property rights and the application of dispute resolution mechanisms such as courts. The politics that ensues is consistent with the state's interest in promoting industrial development. Internationally, the classic case of transboundary pollution is resolved through diplomatic exchange. Because no states, even recipient states, want to entirely curtail such activity, a compromise solution is generally possible, although negotiations may be long and complicated. The politics of reaching such a compromise is not unlike that in resolving transboundary problems with goods in trade or even with threats of military aggression. Terms of trade are set and compensation arranged, or a line is drawn and territory divided.

In these two situations – domestic mediation of conflicting interests, and diplomatic compromise among polluting and polluted states – states can continue to promote industrial development such that economic growth becomes the compromise solution. Much as any organization, profit-making or non-profit, must manage conflicting interests internal and external to the organization, the state engages in such traditional forms of environmental politics simply to do what all states must do to ensure – or hope for – a growing economy and a strong defence. The politics involved are thus consistent with that of the traditional realms – social and economic development and national security. The politics are also consistent with biophysical conditions where pollutants are dispersed and assimilated, where exhausted soils can lie fallow and recover, where the disturbance of one step in the food web is compensated elsewhere in the web, where habitats can be preserved by excluding human use, and so forth. In short, these traditional national and international politics are sufficient when resources and waste sinks are large enough to accommodate all economic activity. When they are not, however, a new form of politics necessarily emerges.

In part, the new politics addresses problems that are truly global in scale and impact – for example, global warming and ozone depletion. But many others are 'global' because they defy prevailing assumptions about resource and waste sink availability – namely, that they are either infinite or that, when the prices are right, new technologies will alleviate shortages. Thus, the Great Lakes cannot assimilate persistent toxic substances, African range states cannot export unlimited quantities of ivory, and the Antarctic cannot withstand conventional mining operations.

Biophysically, then, the global environmental crisis is characterized by the lack of resource and waste sink capability. Socially, it is characterized by the unavoidable involvement of a wide range of stakeholders. When pollutants disperse and bioaccumulate, or when high-income consumption in the North threatens distant habitats and peoples in the South, stakeholders expand beyond the respective states and their industrial partners. In the Great Lakes, for example, stakeholders include all those exposed to persistent bioaccumulative substances, but especially those who eat at the top of the food web and their offspring. In Africa, stakeholders are not just the game wardens and authorized hunters, but all residents who depend, directly or indirectly, on wildlife.

The politics of the global environmental crisis is, therefore, fundamentally different from traditional politics, environmental or otherwise. A state can no longer simply mediate among conflicting interests at home, nor can it strike compromise settlements abroad. When resources are near exhaustion and waste sinks near capacity, when environmental effects are unpredictable and irreversible, when further economic growth contributes more to the problem than to the solution, there are no compromise solutions in the traditional sense of politics.[2] Where a compromise solution is inadequate and long-term effects are unavoidable, the politics of global environmental crisis is necessarily a politics that connects biophysical conditions and engages a wide range of actors. It is a politics that defies traditional, compromise solutions among states and their industrial partners. And it is a politics that pits state industrial interests against community and ecosystem interests.

These characteristics of the global ecological crisis also challenge traditional science and state reliance on that science to solve what are not strictly scientific or technical issues. Traditional, state-centred political and scientific problem-solving mechanisms have been atomistic – single-species, single-chemical, single-medium – not systemic and holistic in their approaches. Environmental problems that are non-linear and unpredictable in their effects and, especially, that create irreversibilities, overwhelm such managerial approaches and render scientific debates intractable.[3] For example, neither the Canadian nor the US governments – let alone municipalities or states or provinces – have been able to effectively deal with toxics in the Great Lakes. Neither CITES nor range states have been able to handle the rapid and geographically differentiated declines of elephant populations, where the best scientific management schemes could not accommodate Northern demand or the needs of resident peoples. Minerals exploitation in Antarctica was unlike all previous 'peaceful uses', and no decision-making system could expunge the politics, let alone safeguard a fragile ecosystem. And, as historically conceived, the United Nations system, stymied by the twin crises of global pollution and poverty, has effected little change, even with such a massive effort as UNCED.

The global ecological crisis, the need for a new politics, and the inadequacies of scientific management challenge the capacities of the traditional state-centric system and thus open a critical niche in world politics. It is a niche not satisfactorily filled by governments or their intergovernmental organizations. It is a niche NGOs have simultaneously helped create and fill to influence both state and society. And it is a niche that makes critical connections among many of the characteristics of the global ecological crisis and in ways quite unlike that found in traditional politics or social movements. We capture the connections international environmental NGOs make in the term, *translational linkages*, in which there are both static and dynamic dimensions.

TRANSLATIONAL LINKAGES: THE STATIC DIMENSIONS

NGO interactions in world environmental politics can be viewed as the construction of linkages on two dimensions: one dimension connects the biophysical to the political, and the other connects the local to the global. First, as argued, biophysical conditions become increasingly politicized as environmental crisis creates conflicts that transcend traditional boundaries and exceeds institutional capacities. Second, global solutions require local approaches when global environmental crisis results from both the aggregation of local resource decisions and from the impact of the global political economy on local communities.[4] Moreover, to the extent that local approaches approximate the conditions for sustainable economies, global solutions must necessarily be based locally.[5] But because local approaches cannot escape from global processes, economic or political, solutions will require connections between the two levels to enable local tailoring and encourage locally indigenous experiments.

Each of these two dimensions – the biophysical/political and the local/global – constitutes a continuum, the ends of which locate the traditional realms of non-governmental intervention. Scientific NGOs, such as SCAR in Antarctica or IUCN in Africa, have concentrated on the biophysical with their scientific research and meetings. Animal rights NGOs have emphasized the political, especially consciousness-raising to change consumer behaviour and national policies. International development NGOs have encouraged local solutions with in-country projects in Southern countries. United Nations accredited NGOs have concentrated on changing the international system.

Emphases on the ends of these continua, on science or national policy or local projects or the UN system, all play a part and account for what international environmental NGOs do. In this book, however, we argue that NGOs' most significant impact on world environmental politics does not derive so much from such single-focus efforts. To the extent that an NGO (or coalition or network) invests primarily in strengthening one end of these two dimensions, its translational influence and effectiveness will be minimal. And as a result, it will violate a tenet of ecological crisis: complex interconnections require systems approaches, not changes in pieces of the system. It will also violate what, in an increasingly globalized world, appears to be a tenet of the political economy of environmental degradation: degradative forces occur at all levels and thus must be addressed simultaneously.

To demonstrate an undue emphasis on any one end of these continua, in Chapter 6 we saw how in the Antarctic, scientific research was consistent with the original Antarctic Treaty System (ATS) principle of peaceful use. The emphasis was on the biophysical end. But when minerals exploitation and biospheric impact were added to the agenda, science was not enough to maintain the regime and adapt at the same time. Because environmental problems pit economic interests against overarching concerns for sustainability and ecosystem integrity, they are inherently political problems. Thus, to promote regime change (as opposed to maintenance, which scientific research did so well), the environmental NGOs supplanted the scientific NGOs. Only the environmental NGOs combined their attention to the biophysical (the two research bases) with international politics (lobbying at both the national and international levels and raising public concern generally). Moreover, only the NGOs could transcend the states' preoccupations with territorial claims and the division of rewards and responsibilities of mineral exploitation. Through these linkages the international environmental NGOs, unlike the strictly scientific NGOs and the states themselves, in effect, set the example that transcended traditional solutions by politicizing the biophysical.[6]

With respect to the emphasis on the local end, in Chapter 5 we saw how mainstream conservation NGOs could invest in community empowerment projects, yet be stymied by the international move to shut down the ivory trade. International development projects in general tend to focus on communities, while ignoring the larger forces resulting in poverty and environmental decline. Examples of this abound in the international development literature.[7]

With respect to the emphasis on the political end, in Chapter 4 we saw how, when GLU joined with the US and Canadian governments to redirect responsibility for monitoring GLWQA implementation to the respective national governments, their transnational allies at the supranational level (the IJC) and at the local level (in the form of the RAPS) were weakened. Similarly, Chapter 5 showed how international conservation which emphasizes state-run, coercive policies that exclude resident peoples was vulnerable to both domestic and international forces.

With respect to the emphasis on the global end, in Chapter 7 we saw how many NGOs invested in strengthening the UNCED process, but neglected the positions of NGOs operating at the grass-roots level. Those NGOs that accepted their assigned role of information disseminator and promoter got visibility and some degree of access. But they also alienated many NGOs, especially the grass-roots and social change NGOs, who were largely excluded and whose views did not get translated to the international level. By contrast, at Rio itself, many NGOs directed their efforts toward strengthening the global NGO community, as opposed to strengthening the state system through the UNCED process. One mechanism was the so-called alternative treaty writing process, a complex and trying effort to write statements of NGO self-commitment. To do this, both vertical and horizontal connections had to be made, connections that, in the implementation phase of the UNCED process, may prove critical. Thus, it appears that those NGO activities in UNCED that conformed to the states' conception of appropriate activity – traditional, essentially national politics yet in an international forum – were most visible but probably least effective in translating local and biophysical needs to the global level. The real translational linkages began to take shape apart from the official process, the concrete effects of which will only be seen over the long term.

Making linkages along the two translational dimensions – biophysical–political, local–global – is essential to change in world environmental politics and, hence, to effective NGO intervention, for three reasons. First, as argued in Part I, traditional diplomacy and international development are ill-equipped to deal with the complexities of global environmental crisis. Moreover, as argued in Chapters 4 and 5, those who best understand the biophysical realities – ecologists, indigenous peoples and others who interact closely with natural systems – that is, those who understand the problems in an ecosystemic framework (rather than diplomatic or economic terms), are not necessarily those inclined to challenge state-sponsored research funding sources or to act politically in the international system. With respect to scientists and ocean dumping, two environmental activists argue:

> that scientists are a lobby group in their own right. This is because almost all scientists involved in the development of the 'prediction' and 'accept-able damage' policy are based in marine science laboratories; and these laboratories rely heavily upon government funding and contracts . . . . Thus, it is our perception that the strongest defenders in the pro-dumping lobby have been not government regulators or industrialists intent on cheap options, but marine scientists with a lifelong record of involvement in dumping programmes.[8]

NGOs often conduct their own scientific research in part to counter this tendency toward state dependency within the scientific community.

With respect to indigenous peoples and others such as farmers, pastoralists, and fishers, the knowledge that supports their livelihoods is grounded in the characteristics of the natural resources. But this knowledge is not readily transferred to the dominant political systems. Consequently, many NGOs can operate at the grass-roots level or link up with those who do. This helps them ground their activities in such knowledge.

In sum, one reason linkages are essential is that, when those who interact closely with natural systems do not translate their knowledge to the political realm, others must. A critical feature of NGO intervention is thus to link the essential knowledge base (scientific and earth-centred) to the world of politics, to translate biophysical needs into choices a wide range of actors can make at many levels of decision making. If NGOs do not make these linkages and translations, they may still operate effectively as lobbyists or green parties, as discussed in Chapter 3, but they are not likely to arrest those processes that have local and global elements and are multi-level and multi-actor and that place a premium on specialized forms of knowledge, which, arguably, characterize the global ecological crisis.

The second reason these linkages are essential to effective NGO intervention is that states are constituted primarily to defend borders and to promote industrial development.[9] States are either too large to meet local environmental needs (whether at home or in their foreign aid policies) or too small to address global issues. They operate to promote and to maintain the integrity of the state, for which the pursuit of industrial development is a necessary condition. As economic prosperity and competitiveness become strategic assets for states, whether to obtain capital or to enter global markets, industrial development becomes coterminous with the maintenance of the integrity of the state.[10]

Consequently, environmental protection at the national or international level is always a subordinate goal to industrial development. In fact, environmental protection is readily subsumed under industrial development by the goal of economic growth, which appears to solve both problems.

To illustrate the reluctance of states to curtail industrial development for environmental protection, in Chapter 4 we saw that a basin-wide governance structure was needed to address a basin-wide pollution problem in the Great Lakes, one that crossed national boundaries. Neither the federal governments of the United States or of Canada were able to do this alone. Even their jointly created IJC was ineffective, as it lacked the necessary authority or ability to act at the local level, where remediation had to take place. The NGOs, however, could facilitate the dual need for a global approach (in this case, regional and binational) and local application by interacting with both the IJC and local communities, in effect bypassing the states. They could also work with the international organization (the IJC) trading on its binational legitimacy while advocating political positions impossible for the IJC to take on its own.

Similarly, we saw in Chapter 5 that parties to CITES may formally ascribe to the norms and principles of conservation and regulation in the wildlife trade regime, but most countries do little to monitor such trade, let alone change consumption patterns. The treaty helps states coordinate interstate commerce but, in the face of tremendous demand, demand that can over-whelm weak institutions, vulnerable peoples and fragile ecosystems and can even destroy the commerce itself, only change at the consumer level could truly protect the threatened species. It was left to NGOs via a global trade ban to, in effect, assimilate at the consumer level the regime's norms. NGOs thus linked consumer behaviour to international commercial patterns to compel states to act.

This is not to say that states neglect environmental issues altogether. They do write treaties, sign agreements, and attend international meetings. But they are often negligent in their implementation.[11] As seen in the ivory case of Chapter 5, CITES members regularly failed to pay their dues, to turn in national reports, and to impose sanctions on violators. Similarly, in Chapter 6 we saw that ATS members failed to meet environmental standards for their research stations that they themselves had set. The ocean dumping regime experiences the same kind of problems.[12]

It is the failure of states to comply with their own agreements that in part creates a niche that transcends state interests. NGOs find opportunities in this niche to, in effect, implement the treaties, but in ways that are consistent with their own environmental and political agenda. Unlike states and their inter-national organizations, NGOs are not bound by national boundaries. They are accountable not to an electorate but only to their membership and, then, only insofar as membership and revenues are maintained. As argued in Chapter 5, in this capacity, NGOs do not have to be nice to anyone. They can be, and often are, in the business of monitoring, exposing, criticizing, and condemning. They need not compromise on either ecological or ethical principles – or, at least, they need do so much less than states, for which the essence of maintaining good relations is, indeed, compromise and for which industrial growth is central.

The third reason these linkages are essential to effective NGO intervention is that, although local problems require local solutions, they also require an enabling political environment.[13] When states do not provide that enabling environment by themselves, their international agreements can. To do so, however, requires that the agreements embody a vision and a set of principles consonant with local needs, not state needs. Linkages, then, are needed from the local to the global to set the terms of debate and to educate the state about its enabling role. Such a role is quite different from that performed by states in promoting national defence or industrial development.

To elucidate, in the Great Lakes case, the international NGO, GLU, first emerged out of a basin-wide local initiative, then effectively 'adopted' the GLWQA as its own to mobilize public support and to hold the respective states to the agreement. Both countries appeared to prefer to ignore or weaken the agreement, but GLU advanced, among other things, the institutionalization of the RAPs. In the wildlife trade case, historically the NGOs went beyond prodding states to, effectively, create and, now, maintain the endangered species trade regime. And, importantly, NGOs link the regime's principles of trade to the local level by operating at both levels. In the UNCED case, although many NGOs diverted their energies to supporting the UNCED process for its own sake, a few used the process to get language in the agreements to address their local needs, to challenge the prerogatives of, for example, the Bretton Woods institutions and the GATT, and to press their governments to implement policies consistent with the spirit of UNCED and with the premises (if not the conclusions) of the Brundtland report. In the follow-up to Rio, NGOs worldwide have embraced the concept of sustainable development and are pushing their governments not just to revise environmental and development policies, but to create the conditions under which local actors including the NGOs themselves can implement locally based programmes.

In short, NGOs, by politicizing the biophysical and linking the local and the global, are tugging and pulling at states. Although some of what they do is entirely separate from state behaviour, much has the effect of setting the conditions under which states will act or, maybe more precisely, react. And these are conditions states are not disposed to create on their own. Despite the diversity within the NGO community (see Chapter 1), overall NGOs seem to be moving in a direction where the tensions between the biophysical and the political as well as between the local and the global are being increasingly bridged.

TRANSLATIONAL LINKAGES: THE DYNAMIC DIMENSIONS

The biophysical/political and local/global framework provides a snapshot of contemporary NGO interactions. It is, however, necessarily static. It does not reveal the mechanisms of change, especially those associated with bio-physical and social change. Two additional concepts, then, are helpful in accounting for change: institutional transformation and social learning.

Institutional transformation refers to the changes in organizations and regimes in response to environmental decline.[14] Although our sample size is limited to four case studies and the path of transformation varies with the regime, patterns can be discerned. In the Great Lakes case, the biophysical conditions changed from direct water use for industrial and residential development, to sewage problems, to persistent toxic chemicals. Corre-spondingly, the Great Lakes water regime evolved from a primarily water projects administration to a sewage clean-up agency and, now, to a regime with the much thornier task of eliminating and preventing persistent, bio-accumulative, toxic chemicals. The ATS has undergone a similar trans-formation from scientific research to minerals exploitation to the estab-lishment of a 'World Park'. CITES appears to be evolving from a loosely coordinated trade regulation regime to a prohibition regime in which tempor-ary moratoria are common and trade cartels with strict limits on production are established by producer countries. Other studies reveal similar trends. The tropical timber and whaling regimes, for example, are transforming from a strict commodity focus to a conservation focus.[15] The ocean-dumping regime is moving from an acceptable damage approach to a precautionary approach.[16] In all cases, although institutional transformation comes about formally through conventional diplomacy and the writing of international laws, in practice such transformation can be attributed in many cases to the 'norm enforcement' carried on by NGOs.[17]

The second concept to account for change in world environmental politics is social learning. As argued in Chapter 3, the NGO contribution to social learning lies in the unique role NGOs play in changing citizens' relationship with traditional politics. This contribution transforms traditional politics altogether – a qualitative change in the very way politics is perceived and practised. In part, the transformation occurs by enhancing environmental awareness. But we have found in this book that the distinctive NGO contribu-tion is more than educational. It is politicizing the biophysical and linking the local and global. Although the construction of the translational linkages is routed in part through states, much of it effectively bypasses the existing political system to transform society directly. NGOs contribute to societal transformation by framing issues, building communities, and setting examples.

First, NGOs are increasingly prominent forces in framing environmental issues. They help establish a common language and, sometimes, common world-views. Indeed, the history of international environmental politics shows that new ideas have not come from governments or even designated international organizations, but from environmental lobbies and activist groups. It was IUCN, for example, that coined the term 'sustainable de-velopment' in its 1980 World Conservation Strategy, which eventually became the conceptual basis of the Brundtland Report and the entire UNCED process.[18] It was Greenpeace and the Sierra Club that introduced the concepts of zero discharge and pollution prevention in the Great Lakes area. It was Greenpeace in the case of Antarctica that was instrumental in making the idea of a world park acceptable. It was Southern NGOs, in particular the Third World Network, that, in the UNCED process, put environment and development questions in terms of South–North equity.

Second, international environmental NGOs contribute to societal transformation through community development. Some NGOs organize communities at a local level, where they involve citizens in concrete projects. Others build coalitions among communities and across regions and nations to strengthen those communities and to make them more autonomous *vis-à-vis* existing political structures. Such community-focused efforts represent a proactive approach to social transformation. They develop community capacity to design self-reliant economies and to resist intrusive political and economic forces. This approach replaces the traditional political approach which sees the community as a mobilizable constituency for state-defined political and economic purposes.

Community development, as many NGOs are discovering, must take place simultaneously at more than one level. NGOs which operate only at the community level ignore the larger forces impinging on local self-reliance, just as those which operate only at the international level ignore the local component of global processes. Effective NGO intervention makes linkages between the two levels, thus framing the issue as one that is not singularly local nor singularly global. With such NGO-constructed linkages, actors at all levels begin to realize and act on the interconnections and begin to understand the local in terms of the global, and vice versa. Such upstreaming, making explicit political and economic connections from the local to the national and international,[19] accounts for much of the distinctive contribution of environmental NGOs and constitutes a process of social learning.

An example of the community-building function can be found in the Great Lakes case. NGOs such as the National Wildlife Federation or Pollution Probe Canada work directly with local communities to devise pollution control strategies. Moreover, these and other NGOs form horizontal coalitions and strengthen each other by mobilizing public support, commissioning studies, and monitoring private and public compliance. Finally, the coalitions operate vertically by linking their communities to the international level via, for example, the International Joint Commission.

The third way in which international environmental NGOs contribute to societal transformation is by setting examples and substituting for governmental action. Instead of calling for action or mobilizing citizens to put pressure on governments, NGOs often just do the work themselves. In the case of Antarctica, Greenpeace had its own research station where it demonstrated by example how such a station can be run in an environmentally benign way. In wildlife trade, TRAFFIC and other NGOs assume a significant share of what would otherwise be a primary state or intergovernmental function in trade management – monitoring. In the UNCED process, Environmental Liaison Centre International, Friends of the Earth, and others unhappy with the way NGOs were fed into the UNCED process, set up an NGO process parallel to both UNCED and the Global Forum in Rio. In sum, by substituting for governmental action and setting examples with concrete activities, environmental NGOs engage in creative and innovative learning processes whose results come to affect society as a whole.

Institutional transformation and social learning help account for the dynamic dimensions of NGO interactions. By supplementing, replacing, bypassing, and, sometimes, even substituting for traditional politics, NGOs are increasingly picking up where governmental action stops – or has yet to begin. Increasingly independent of traditional politics, NGOs can engage in the new politics associated with global ecological crisis. They can stake out a claim on those environmental issues unresolvable by traditional politics and build their own, often unique, bargaining assets to negotiate with other international actors. In so doing, NGOs do not merely lobby and persuade and provide information. Rather, they act as agents of social learning by linking the biophysical conditions with political concerns while simultaneously acting locally and globally. They frame the issue as both local and global, such that actors at all levels begin to understand the local in terms of the global and vice versa – a process that, indeed, is one of social learning.

Of the two dynamic dimensions, institutional change is generally most visible and, not surprisingly, most prominent in the literature. By its very conceptual nature, social learning is difficult to document. Nevertheless, the case studies reveal much about such learning, especially with respect to the process of translating issues from an ecological understanding to a political one and from concerns about global environmental trends to their meaning for local politics and economic development. That is, whereas knowledge of ecological relationships is widely available in scientific journals and conferences, it takes a transformative imagination to interpret the political implications of that understanding, an imagination generally not held by scientists or indigenous peoples or by states or corporations, at least not in their customary roles. The transformation spurred by NGOs in their role in world politics, therefore, is not a transformation of the entire international system, at least not in the time frame of this study. Rather, it is a transformation through social learning that results from translating ecological knowledge and local–global relationships.

PRACTICE

We have argued that if international environmental NGOs are significant contributors to institutional transformation and social learning, a key element of their contribution is the construction of effective translational linkages. Consequently, NGO effectiveness can be evaluated in terms of movement toward such constructions and away from traditional politics where the state promotes industrial development through domestic mediation and diplomatic compromise. Such NGO practice can be improved to the extent NGOs consciously develop their capacity to promote such linkages. Four practical considerations follow.

Creating the linkages

The notion of translational linkages might be interpreted to mean that every environmental NGO with international pretensions (or, even, international concerns) must deal simultaneously with village leaders and the United Nations or the European Parliament. Our case studies suggest the contrary. Biophysical–political linkages and local–global linkages can be multi-tiered. If there is a prevailing feature of NGO relations at the international level, it is that coalitions and networks are critical to do the work.[20] But as shown in Chapter 2, it is work that is not simply power-building or information-sharing, the usual reasons for domestic coalitions and networks. Rather, it is work that crosses levels of NGO activity and connects specialized NGO needs with generalized NGO access.

Thus, in the Great Lakes case, we saw that individuals and NGOs, small and large, found a compelling need to form a basin-wide organization, not primarily to gain more clout in Washington and Ottawa, but to establish a basin-wide presence that drew first and foremost on local needs across the basin. So, although many of the local linkages were left to GLU member organizations, GLU itself concentrated its efforts on the transboundary and international (IJC) realms, gaining access that local groups, by themselves, could not. In the ivory case, we saw that networks of trade-monitoring NGOs worked with local NGOs, traders, and the CITES secretariat. NGO monitors could not inspect every wildlife shipment, but they could compare export and import data, an activity governments tended to avoid, and report to relevant national and international agencies. In the Antarctic case, one NGO, Greenpeace with its worldwide operations, was instrumental in promoting the idea of an Antarctic World Park. But a host of other NGOs, many operating through the ASOC coalition, was critical to mobilizing domestic support in France, Australia and elsewhere. In the UNCED case, throughout the PrepComs and in Rio itself, we saw networks form. This was particularly pronounced at the regional level, suggesting that national-level organization (the predominant level, still) was insufficient and that, quite possibly, truly global networks were too cumbersome and of doubtful utility. The legitimacy of these networks depend critically on their ability to serve local needs.[21]

To be effective builders of translational linkages, we see that single NGOs do not have to make all the connections. Some large, well-financed NGOs can conduct on-the-ground projects and still hobnob with World Bank executives. But most can be effective via networks and coalitions. The effectiveness test for these coalitions is not, as noted, the apparent clout according to budget or membership nor the quantity of information exchanged. Rather, it is the extent to which, first, biophysical realities are translated effectively into political action and, second, local needs are transferred to international decision-making (and then to national decision-making).

Building credibility as a special representative of environmental interests

Government officials, especially popularly elected ones, do not want to hear claims that NGOs better represent environmental interests or, for that matter, public health or women's or indigenous rights or development interests. Although the point is highly debatable and raises fundamental questions of representation and democracy, to make credible claims to transnational environmental values, NGOs must find means of demonstrating their claim in a manner which is recognizable by other international actors. It is not enough to cite public opinion polls that purport to show that NGOs are more trusted than governments or corporations.

So, for example, to make a claim on environmental values in the Antarctic, Greenpeace did not merely hang another banner at an ATS meeting. It came as close to being an ATS member as a non-state actor could, by establishing a bona fide research base (something many states are unable to do). Moreover, it set the example for environmentally sound research by generating much of its own energy, disposing of its wastes properly, and performing its own environmental impact statements. Similarly, in the endangered species trade regime, WWF and others offer something that many states cannot provide themselves: rules and procedures for wildlife trade, monitoring beyond one's own borders, and some degree of enforcement. In the Great Lakes case, GLU conducted a basin-wide (not just American or Canadian or, even, 'binational') tour to solicit public input on the renegotiation of the GLWQA. A basin-wide public view was something neither party to the bilateral negotiations could claim. And in the UNCED process, those NGOs that rejected the role of information disseminator did more than complain and protest. They engaged in a quasi-institution-building process to articulate an alternative vision of sustainable development. And, possibly most significantly, they began to coordinate their own activities; activities, they argued, the states would eventually follow.

In short, NGOs are effective agents of change to the extent that they operate independently of states and do what states tend not to do. By performing such functions, rather than lobbying others to perform them, NGOs can make a credible claim on legitimate environmental representation, and they can do so in ways governments and corporations cannot.

Organizing transnationally

Despite the natural proclivity to organize nationally, many NGOs may be more effective eschewing national identities as an organizing concept. For example, as demonstrated in Chapter 4, GLU's organizational boundaries more closely fit the issue at hand (namely, ecosystem management) than the boundaries of any existing political entity. WWF-Int. operates worldwide but, with IUCN, concentrates much of its activities in Geneva, near the CITES headquarters, and it sets up TRAFFIC offices in key countries involved in wildlife trade. It certainly works with national governments but it organizes to fit the trade.

A potential application of this prescription can be found in the follow-up to UNCED. Just because states are charged with implementation does not mean the parallel NGO organizing need be national. In fact, in North America, the most important follow-through will be, arguably, that of the United States, and not just for US citizens but for Canadians and Mexicans as well. Thus, although seemingly paradoxical, the NGO organization that emerges to promote US implementation could well involve nationals of neighbouring states and, to push the argument to its limit, to nationals of any state that is significantly affected by US foreign economic policy. Such an organization would be able to provide a transnational channel of influence on a country that, in its own right, is the world's foremost purveyor of transnational economic activities.

Interacting with international organizations

As noted, international organizations seek NGOs as constituents when their mandates are broad and their authority limited. And, as argued in Chapter 7, NGOs must be wary of being co-opted. But to put the relationship in bargaining terms, as argued in Chapter 2, NGOs can trade on their ability to provide a constituency by demanding the means of creating linkages from the local to the international (and back to the national). Thus, NGOs will have effective transnational relations with international organizations to the extent that the NGOs assert their autonomy as independent actors and to the extent they can credibly represent the interests of local communities seeking sustainable economies.

CONCLUSION

In this book we have attempted to come to grips with the rapidly emerging and changing global NGO phenomenon. We began our inquiry by asking, for example, what it means when 30,000 NGO representatives converge at the Global Forum in Rio, or when global donor agencies turn to NGOs for the implementation of their environment and development programmes or when representatives of environment and development NGOs sit down with business leaders to negotiate acceptable pollution levels. In our attempt to address such questions we have begun in this study to articulate the nature of the growing role that environmental NGOs have come to play in transforming world environmental politics.

We conclude that international environmental NGOs make their most distinctive contribution by going beyond traditional politics, that is, beyond state-oriented practices designed to ameliorate the side effects of industrial development. NGOs make their contribution when they translate biophysical change under conditions of global ecological crisis into political change and do so at both the local and global levels.

Lobbying one's own government or educating the public at home or abroad does not exploit this distinctiveness. Rather, this study suggests that NGOs are most effective at the international level to the extent they exploit transnational opportunities. And, whereas NGOs of all kinds – human rights, women's, public health, and so forth – also exploit transnational linkages, environmental NGOs inject scientific and earth-centred concerns into political and economic situations which would otherwise relegate such concerns to the margins. In this *translational* mode, environmental NGOs transform politics by redefining what constitutes its subject matter.

In this study, we have not addressed the full range of NGO relations or the variability in their interventions or questions about NGOs' own organization. In many ways, we have only begun to lay out a research agenda, one that can use the framework developed here to test, build, refine, and, where necessary, discard those elements of NGO relations that do not fit. A comprehensive study of NGO translational relations must go further. It must examine NGO relations, both in the South and from the Southern perspective. It must seek cases that allow comparative analysis by holding, say, the institutional context of the biophysical conditions constant, and varying the nature of the NGO and its intervention.

A more comprehensive study must also examine the tensions and conflicts within the NGO community and within the organizations themselves. Although this book has focused primarily on NGOs' external relations, a fuller account would examine intra-NGO dynamics and connect them to the external dynamics. For example, many NGOs are undergoing profound self-examination as they make the shift from narrowly focused campaigns to comprehensive sustainable development programmes. The difficulties this engenders go beyond the familiar organizational conflicts that NGOs experience as voluntary or membership organizations or as foundation- or government-dependent organizations. Global transition to sustainable economy and broad-based social learning has its counterpart in organizational transition and organizational learning. These, in turn, will affect how NGOs translate the biophysical to the political, link the local and the global, bargain with actors at all levels, and promote social environmental learning.

NOTES

1 Three authors working in the tradition of the Brundtland Report and the UNCED process put it even more apocalyptically:

> The horizon may glow with technological opportunities, but the obstacles to sustainability are not mainly technical; they are social, institutional and

political. Given the constraints on social, institutional and political change, no one can rule out a future of progressive ecological collapse. The four Horsemen of the Apocalypse – war, famine, pestilence, and death – are galloping through parts of Africa, Asia, and Latin America, and they will surely remain active, spurred on by increasing poverty and greed, and by policy and institutional failures. Threats to the peace and security of nations from environmental breakdown are increasing at a frightening pace. Conflicts based on climate change, environmental disruption, and water and other resource scarcities could well become endemic in the world of the future.

> (Jim MacNeill, Pieter Winsemius, and Taizo Yakushiji,
> *Beyond Interdependence: The Meshing of the
> World's Economy and the Earth's Ecology*
> (New York: Oxford University Press, 1991) 19–20)

2 See Chapter 5 for a discussion of the biophysical, psychological, organizational, and political determinants of no-compromise solutions.

3 For analysis of the inadequacies of traditional science, see Charles Perrings, 'Reserved Rationality and the Precautionary Principle: Technological Change, Time and Uncertainty in Environmental Decision Making', in Robert Costanza, ed., *Ecological Economics: The Science and Management of Sustainability* (New York: Columbia University Press, 1991). Perrings concludes:

> As our knowledge of the global system increases, so does our uncertainty about the long term implications of present economic activity. Combined with the uncertainty caused by the rapid pace of change in resource use technology, this suggests that the increasing flow of information does not in fact give more complete information. The problem for decision makers does not get easier. Not only is the perceived range and severity of the possible environmental effects of economic activity expanding, so is the gestation period. (164)

4 For arguments regarding the grounding of global solutions at the local level, see Chapters 2, 4, and 5. Notice also that each of these terms is a shorthand expression of a more complex concept. Thus, *local*, as discussed in Chapter 2, is rarely defined in the literature. Similarly, *political*, as discussed here and in Chapters 1 and 3, refers primarily to traditional, nationally oriented politics. The politicization of the biophysical can be seen in the first instance as elevating environmental changes to the political realm, that is, to the traditional forms of political discourse, especially national environmental politics; and, in the second, transforming that realm. It is the transformative aspects that, in an increasingly globalized world, are distinctively transnational components and for which international NGOs are key players.

5 Growing literatures in international development, common property resource regimes, environmental psychology and elsewhere indicate that environmental problems must indeed, be grounded in local conditions. Some authors argue that solutions must be 'tailored' to local conditions, thus implying the primacy of top-down approaches. See, for example, H. Jeffrey Leonard, ed., *Environment and the Poor: Development Strategies for a Common Agenda* (New Brunswick, NJ: Transaction Books). Others speak of local conditions that allow for indigenous approaches yet require a larger – national and international – enabling environment. See, for example, Elinor Ostrom, *Governing the Commons: The Evolution of Institutions for Collective Action* (Cambridge: Cambridge University Press, 1990). For further discussion of the top-down and bottom-up approaches and the need for an analysis of the local dimension of global environmental problems, see Chapter 2.

6 Notice that there is no implied balance within the biophysical/political and local/

global framework. Thus, because there are no human communities in Antarctica, the only relevant dimension in the Antarctic case is the biophysical/political.

7 See, for example, Leonard, *Environment and the Poor*; John Lewis and contributors, *Strengthening the Poor: What Have We Learned?* (New Brunswick, NJ: Transaction Books, 1988).

8 Kevin Stairs and Peter Taylor, 'Non-Governmental Organizations and Legal Protection of the Oceans: A Case Study' in Andrew Hurrell and Benedict Kingsbury, eds., *The International Politics of the Environment: Actors, Interests, and Institutions* (Oxford: Oxford University Press, 1992) 122–3.

9 This is, of course, a highly simplified characterization of states. The important point, however, is that states are largely removed from local conditions, especially resource management conditions, and from the governance structures that often prevail at that level. Borrowing from Migdal, we can distinguish the state from 'the local' in terms of a 'melange' of organizations, state and social. The state is

> an organization, composed of numerous agencies led and coordinated by the state's leadership (executive authority) that has the ability or authority to make and implement the binding rules for all people as well as the parameters of rule making for other social organizations in a given territory, using force if necessary to have its way. (19)

The social organizations, by contrast, cannot use military force but compete with the state for social control: 'The central political and social drama of recent history has been the battle pitting the state and organizations allied with it (often from a particular social class) against other social organizations dotting society's landscape' (27–8). Social organizations include families, neighbourhood groups, clans, clubs, and communities (25) and thus correspond to what we term 'local'. Although Migdal does not address resource questions, let alone the need to tailor resource management schemes to specific ecosystem conditions as we argue here, he does warn analysts about the dangers of dismissing local organizations in favour of state organization:

> It has been far too common in the literature on the Third World to dismiss with a wave of the hand the importance of the local, small organizations with rules different from those of the state. They have seemed so inconsequential, especially to someone who has rarely left the capital city. (36)

(Joel S. Migdal, *Strong Societies and Weak States: State–Society Relations and State Capabilities in the Third World* (Princeton, NJ: Princeton University Press, 1988)

10 Michael E. Porter, a leading analyst of competitive industrial strategy, characterizes states and their national identity primarily in terms of their respective competitive economic advantages in the pursuit of industrial development. That is, competitive states attract competitive industries which in turn make states more competitive.

> The role of the home nation seems to be as strong as or stronger than ever. While globalization of competition might appear to make the nation less important, instead it seems to make it more so. With fewer impediments to trade to shelter uncompetitive domestic firms and industries, the home nation takes on growing significance because it is the source of the skills and technology that underpin competitive advantage.
> (*The Competitive Advantage of Nations* (New York: The Free Press, 1990, 19)

From a southern perspective, Roberto P. Guimarães characterizes Brazil in terms of 'The primacy of economic growth and industrialization over conservation and the rational use of natural resources [which] constitutes, perhaps, the oldest part of ecopolitical ideology in Brazil, dating back to colonial times' (*The Ecopolitics of Development in the Third World: Politics and Environment in Brazil*, (Boulder, CO: Lynne Rienner Publishers, 1991) 159).

11 See, for example, Patricia Birnie, 'International Environmental Law: Its Adequacy for Present and Future Needs' (51–84), and Andrew Hurrell and Benedict Kingsbury, 'Introduction' (1–47), both in Andrew Hurrell and Benedict Kingsbury, eds., *The International Politics of the Environment: Actors, Interests, and Institutions* (Oxford: Oxford University Press, 1992); Lynton K. Caldwell, *International Environmental Policy: Emergence and Dimensions* (Durham, NC: Duke University Press, 1984; 1990).

12 Kevin Stairs and Peter Taylor, 'Non-Governmental Organizations and Legal Protection of the Oceans: A Case Study' in *The International Politics of the Environment*, 116–17.

13 Elinor Ostrom refers to the enabling political environment as 'nested enterprises' (*Governing the Commons: The Evolution of Institutions for Collective Action*, Cambridge: Cambridge University Press, 1990) 101–2. In Migdal's terminology, it may be the 'parameters of rule making' (largely from the state) and the 'manipulation of symbols about how social life should be ordered' (largely from social organizations, many of which are local; see note 9 above) that provides the enabling environment for specific resource management regimes.

14 Oran Young and his colleagues have done the critical research on natural resource regime change. See, for example, Oran R. Young, *International Cooperation: Building Regimes for Natural Resources and the Environment* (Ithaca, NY: Cornell University Press, 1989); Gail Osherenko and Oran R. Young, *The Age of the Arctic: Hot Conflicts and Cold Realities* (Cambridge: Cambridge University Press, 1989); Oran R. Young and Gail Osherenko, *Polar Politics: Creating an Environmental Regime* (Ithaca, NY: Cornell University Press 1993).

15 One might argue that these regime changes are not so much transformations as assimilations of contemporary concerns. The World Bank creates an environment division and then continues to promote unsustainable projects. But the examples given reveal that the norms and principles and rules and procedures are changing. The International Tropical Timber Organization (ITTO), for example, places conservation as its top priority whereas only ten years ago the agreement that led to its creation placed conservation last after a long list of conventional commodity principles. See Thomas Princen, 'From Timber to Forest: The Evolution of the Tropical Timber Trade Regime', typescript, 1993.

16 Kevin Stairs and Peter Taylor, 'Non-Governmental Organizations and Legal Protection of the Oceans: A Case Study' in Andrew Hurrell and Benedict Kingsbury, eds, *The International Politics of the Environment: Actors, Interests, and Institutions* (Oxford: Oxford University Press, 1992) 116–17.

17 We thank Michael Ross for bringing this component of regime change and the role of NGOs to our attention. See Chapter 5 for analysis of one mechanism – bans – by which international norms are developed.

18 International Union for the Conservation of Nature, *The World Conservation Strategy* (Gland: International Union for the Conservation of Nature, 1980).

19 David Korten, *Getting to the 21st Century: Voluntary Action and the Global Agenda* (West Hartford, CT: Kumarian Press, 1990).

20 Bramble and Porter come to a similar conclusion regarding the importance of coalitions and networks in their assessment of the NGO role in international negotiations over multilateral aid, ozone depletion, and tropical timber. Barbara J. Bramble and Gareth Porter, 'Non-Governmental Organizations and the Making of US International Environmental Policy', in Andrew Hurrell and Benedict Kingsbury, eds, *The International Politics of the Environment: Actors, Interests, and Institutions* (Oxford: Oxford University Press, 1992) 313–53.

21   The regional emphasis may parallel the global trend toward regional trading blocs. It may also indicate a general disappointment with fully global efforts such as Environmental Liaison Centre International or the International Facilitating Committee of the Center for Our Common Future (see Chapter 7). More likely, however, it reflects a general reluctance within the NGO community to create such a large-scale organization, even if only a loose network, and, at the same time, a recognition that solving one's own local or national problems is not enough. For example, a New Zealand NGO achieved numerous successes domestically, only to find that the South Pacific generally was suffering a number of serious environmental abuses. At Rio, NGOs from numerous islands, New Zealand and Australia laid the groundwork for a regional organization. Many North American, European, and Japanese NGOs are making similar discoveries and are creating similar organizations.

Bibliography

Acheron, Marilyn, 'Africa: The Last Safari?' *Newsweek*, August 18 (1986): 40–42.

Adgham, Munir, 'Non-Governmental Environmental Organizations in the Gulf of Aqaba-Bordering States: A Current Appraisal', in Environmental Law Institute, *Protecting the Gulf of Aqaba: A Regional Environmental Challenge*, Washington, DC: Environmental Law Institute, 1993.

Alexander, L., and Lynn Hanson, eds, *Antarctic Politics and Marine Resources: Critical Choices for the 1980s*, Kingston, RI: Center for Ocean Management Studies, 1984.

Alper, Joseph, 'Should Heads Keep Rolling in Africa?', *Science* 255 (1992): 1206–7.

Amer, Omnia, 'An Agenda for Cooperation Among Non-Governmental Organizations in the Gulf of Aqaba-Bordering States', in Environmental Law Institute, *Protecting the Gulf of Aqaba: A Regional Environmental Challenge*, Washington, DC: Environmental Law Institute, 1993.

Andersen, Steinar, and Willy Ostreng, eds, *International Resource Management: The Role of Science and Politics*, London: Belhaven Press, 1989.

Antarctic and Southern Ocean Coalition. Press releases: April 30, 1991, and June 14, 1991.

Antartic and Southern Ocean Coalition, 'Reports on the Second and Third Sessions of the XI Antarctic Treaty Special Consultative Meetings', August 28, 1991.

'Antarctic Compromise "Disastrous"', *The Australian*, June 22 (1991).

'Antarctic Environment Protected for 50 Years', *Eugene Register-Guard,* October 5 (1991): 7A.

Antarctica Project Newsletter 1(2) (Summer 1992) 3.

'Anti-logging Priest Killed in Ambush', *Haribon Update*, newsletter of the Haribon Foundation, 7(1) (January–February 1992): 2,3,4–5.

Ashworth, William, *The Late Great Lakes: An Environmental History*, New York: Alfred A. Knopf, 1986.

Auburn, F. M. *Antarctic Law and Politics*, Bloomington: Indiana University Press, 1982.

Aufderheide, Pat, and Bruce Rich, 'Environmental Reform and the Multilateral Banks', *World Policy Journal* 5 (2) (1988): 300–21.

Barbier, Edward B., Joanne C. Burgess, Timothy M. Swanson, and David W. Pearce, *Elephants, Economics and Ivory*, London: Earthscan Publications, 1990.

Barnes, J. *Let's Save Antarctica!* Victoria, Australia: Greenhouse Publications, 1982.

Barkin, David, and Steve Mumme, 'Environmentalists Abroad: Ethical and Policy Implications of Environmental Non-Governmental Organizations in the Third World', Paper prepared for the Third Congress of the International Development Ethics Association, Tegucigalpa, Honduras, June 21–8, 1992.

Beck, P. *The International Politics of Antarctica*, New York: St. Martins Press, 1986.

Beeby, C. 'The Antarctic Treaty System as a Resource Management Mechanism–Nonliving Resource', in Polar Research Board, ed., *Antarctic Treaty System: An Assessment*. Proceedings of a workshop at Beardsmore, South Field Camp, Antarctica, January 7–13, 1985. Washington, DC: National Academy Press, 1986.

Bell, Daniel, *The Coming of Post-Industrial Society. A Venture in Social Forecasting*, New York: Basic Books, 1973.

Benedick, Richard E. *Ozone Diplomacy: New Directions in Safeguarding the Planet*, Cambridge, MA: Harvard University Press, 1991.

Bergin, Anthony, 'The Politics of Antarctic Minerals: The Greening of White Australia', *Australian Journal of Political Science*, vol. 26 (1991): 216–39.

Bernstein, Johannah *et al. PrepCom III: Third Week Synopsis*. NGO strategy group paper, Geneva, 1991.

Birnie, Patricia, 'The Role of International Law in Solving Certain Environmental Conflicts', in John E. Carroll, ed., *International Environmental Diplomacy*, Cambridge: Cambridge University Press, 1988.

Birnie, Patricia. 'International Environmental Law: Its Adequacy for Present and Future Needs', in Andrew Hurrell and Benedict Kingsbury, eds, *The International Politics of the Environment: Actors, Interests, and Institutions*, Oxford: Oxford University Press, 1992: 51–84.

Boardman, Robert, *International Organization and the Conservation of Nature*, Bloomington: Indiana University Press, 1981.

Bodansky, Daniel, 'Scientific Uncertainty and the Precautionary Principle', *Environment* 33(7): 4–5, 43–4.

Bonner, Raymond, 'Crying Wolf Over Elephants: How the International Wildlife Community Got Stampeded into Banning Ivory', *New York Times Magazine*, February 7 (1993): 16–19, 30, 52–3.

Boyer, Barry, 'Ecosystem, Legal System and the Great Lakes Water Quality Agreement', Paper presented at the Annual Meeting of the International Association of Great Lakes Researchers (IAGLR), Buffalo, NY, June 1991.

Bramble, Barbara J. and Gareth Porter, 'Non-governmental Organizations and the Making of US International Environmental Policy', in Andrew Hurrell and Benedict Kingsbury, eds, *The International Politics of the Environment*, Oxford: Oxford University Press, 1992, 313–53.

Brecher, W. Puck, 'CMs [Citizen Movements] and NGOs: Grassroots Perspectives on the Japanese Environmental Gridlock', masters thesis, Asian Studies, University of Michigan, August 31, 1992.

Brewer, Garry, 'Environmental Challenges and Managerial Responses', in Nazli Choucri, ed., *Global Accord: Environmental Challenges and International Responses*, Cambridge, MA: MIT Press, 1993: 281–305.

Browder, John O. 'Development Alternatives for Tropical Rain Forests', in H. Jeffrey Leonard, ed., *Environment and the Poor*, Washington, DC: Overseas Development Council, 1989.

Brown, L. David, and David C. Korten, 'Understanding Voluntary Organizations: Guidelines for Donors', WPS 258, Working Paper, Country Economics Department, Washington, DC: World Bank, September 1989.

Brown, Lester R. *et al. State of the World*, New York: Norton, 1992.

Brown, M. Leann. 'Agenda Setting, Policy Making, and Institutional Learning in an International Setting: The Greening of the European Community', Paper presented at 1992 annual meeting of the International Studies Association, Atlanta, GA.

'Bush Comes Around on Antarctic Mining Ban', *Sydney Morning Herald*, July 5, 1991.

Caldwell, J. R. 'The Effects of Recent Legislative Changes on the Pattern of the World's Trade in Raw Ivory', *TRAFFIC Bulletin* 9 (1): 6–10.

Caldwell, Lynton K. *In Defense of Earth: International Protection of the Biosphere*, Bloomington: Indiana University Press, 1972.

Caldwell, Lynton K. *International Environmental Policy: Emergence and Dimensions*, Durham, NC: Duke University Press, 1984; 1990.

Caldwell, Lynton K. *Between Two Worlds: Science, the Environmental Movement, and Policy Choice*, Cambridge: Cambridge University Press, 1990.

Caldwell, Lynton K. 'Beyond Environmental Diplomacy: The Changing Institutional Structure of International Cooperation', in John E. Carroll, ed., *International Environmental Diplomacy,* Cambridge: Cambridge University Press, 1988: 13–28.

Caldwell, Lynton K., ed, *Perspectives on Ecosystem Management for the Great Lakes: A Reader*. Albany, NY: State University of New York Press, 1988.

Caponera, Dante A. 'Patterns of Cooperation in International Water Law: Principles and Institutions', in Albert Utton and Ludwik A. Teclaff, eds, *Transboundary Resources Law*, Boulder, CO: Westview Press, 1987: 1–27.

Carlson, D. and C. Comstock, eds, *Citizen Summitry: Keeping the Peace When it Matters Too Much to Be Left to Politicians*, Los Angeles, CA: J.P. Tarcher, 1986.

Carroll, John E. 'Differences in the Environmental Regulatory Climate of Canada and the United States', *Canadian Water Resource Journal* 4 (1979): 16–25.

Carroll, John E. *Environmental Diplomacy*, Ann Arbor, MI: University of Michigan Press, 1983.

Carroll, John E. *International Environmental Diplomacy; The Management and Resolution of Transfrontier Environmental Problems*, Cambridge: Cambridge University Press, 1989; 1990.

Cater, Nick, 'Preserving the Pachyderm', *Africa Report*, November/December (1989): 45–6, 48–9.

Center for Investigative Reporting and Bill Moyers, *Global Dumping Ground*, Washington, DC: Seven Locks Press, 1990.

Chadwick, Douglas H. 'Elephants – Out of Time, Out of Space', *National Geographic* 179 (5): 2–49.

Chatterjee, P. and M. Finger, *The Earth Brokers: Power, Politics and World Development*, London: Routledge, 1994.

Chen, Lung-Chu, *Contemporary International Law: A Policy-Oriented Perspective*, New Haven, CT: Yale University Press, 1989.

Christie, W. J., M. Becker, J. W. Cowden, and J. R. Vallentyne, 'Managing the Great Lakes Basin as a Home', *Journal of Great Lakes Research* 12 (1986): 2–17.

CITES, Fourteenth Annual Report of the Secretariat, Lausanne, Switzerland, (January 1–December 31, 1989).

CITES *Proceedings*, Lausanne, Switzerland, October 8–20, 1989.

CITES Secretariat, 'Interpretation and Implementation of the Convention: Trade in Ivory from African Elephants, Operation of the Ivory Trade Control System', Document 7.21, Lausanne, Switzerland, 9–20 October 1989: 1–20.

CITES Secretariat, 'Interpretation and Implementation of the Convention: Trade in Ivory from African Elephants, Strengthening of the Ivory Trade Control System', Document 7.23, Lausanne, Switzerland, 9–20 October 1989: 744–60.

CITES, Seventh Meeting of the Conference of the Parties, Lausanne, Switzerland. List of Participants. Part. 7.2. 1989.

Clark, Margaret L., and John S. Dryzek, 'The Inuit Circumpolar Conference as an International Nongovernmental Actor.' Paper prepared for the Arctic Policy Conference at McGill University, Canada, September 19–21, 1985.

Colborn, Theodora E., A. Davidson, S. N. Green, R. A. Hodge, C. I. Jackson, R. A. Liroff, *Great Lakes Great Legacy?* Washington, DC: Conservation Foundation; Ottawa, Ontario: Institute for Research on Public Policy, 1990.

Colchester, Marcus, 'International Tropical Timber Organization', *The Ecologist* 20 (5): 166–73.

Cook, Elizabeth, 'Global Environmental Advocacy: Citizen Activism in Protecting the Ozone Layer', *Ambio* 19(6–7) (October 1990): 334–7.

Costanza, Robert, ed., *Ecological Economics: The Science and Management of Sustainability*, New York: Columbia University Press, 1991.

Craig, Gordon A., and Alexander L. George, *Force and Statecraft: Diplomatic Problems of Our Time*, New York: Oxford University Press, 1983; 1990.

Daly, Herman, *Steady-State Economics*, Covela, CA: Island Press, 1991.

Dawkins, Kristin, 'Sharing Rights and Responsibilities for the Environment. Assessing Potential Roles for Non-Governmental Organizations in International Decision-Making.' Unpublished masters thesis, MIT, Cambridge, MA, 1991.

DeBenedetti, Charles, *An American Ordeal. The Antiwar Movement in the Vietnam Era*, Syracuse, NY: Syracuse University Press, 1990.

Deudney, Daniel, 'Global Environmental Rescue and the Emergence of World Domestic Politics', Paper presented at the International Studies Association Annual Meeting, Vancouver, BC, March, 1991.

Dichter, Thomas W. 'The Changing World of Northern NGOs.' in John Lewis, ed., *Strengthening the Poor*, New Brunswick, NJ: Transaction Books, 1988.

Douglas-Hamilton, Dr I. WWF Project 3882, EEC/WWF African Elephant Programme. Activities Report. Nairobi, Kenya. Reporting period: February 15, 1989–March 30, 1990.

Dunlap, Riley, 'Public Opinion in the 1980s. Clear Consensus, Ambiguous Commitment', *Environment* 33 (1991): 10–15, 32–7.

Dunlap, Riley, and Angela Mertig, *American Environmentalism: The US Environmental Movement, 1970–1990*, Washington, DC: Taylor and Francis, 1992.

Durning, Alan Thein, 'Native Americans Stand Their Ground', *Worldwatch*, November/December (1991): 10–17.

Durning, Alan Thein, 'Action at the Grassroots: Fighting Poverty and Environmental Decline.' Worldwatch paper no. 88. Washington, DC: Worldwatch Institute, 1989.

Dworsky, Leonard, B. 'Changes in Management: International Great Lakes', in *A New Agenda for the Management of the Great Lakes*, Ithaca, NY: Cornell University Press, 1990.

Dworsky, Leonard B. 'The Great Lakes: 1955–1985', in Lynton K. Caldwell, ed., *Perspectives on Ecosystem Management for the Great Lakes*, Albany, NY: State University of New York Press, 1988: 59–113.

Dworsky, Leonard B., and David Allee, 'An Agenda for the Management of the Great Lakes on a Long Term Ecosystem Basis', American Water Resources Association, 1989.

Dworsky, Leonard B. and Charles F. Swezey *The Great Lakes of the United States and Canada: A Reader on Management Improvement Strategies*, Ithaca, NY: Cornell University Press, 1974. Based on the report of the Canada-United States Inter-University Seminar (CUSIS), 'A Proposal for Improving the Management of the Great Lakes of the United States and Canada', 1974.

Earth Summit News. Econet, electronic news, topic 744, San Francisco, CA: Institute for Global Communications, May 26, 1993.

'East/West Non-Governmental Organizations Action Plan for the Ecological Reconstruction of Central and Eastern Europe'. Statement of the Vienna Conference on the Ecological Reconstruction of Central and Eastern Europe, November 15–17, 1992.

ECO. Various issues, especially vol. 69, Bonn, Germany, April 15 1991; and vol. 70, Madrid, Spain, April 22–30 1991. Nos. 1, 2, and 3.

Economist, The. 'Saving the Elephant', 1 July 1989: 15–17.

Ekins, Paul, *A New World Order: Grassroots Movements for Global Change*, London: Routledge, Chapman and Hall, 1992.

El-Kholy, Osama and Mostafa Tolba, eds, *The World Environment 1972–1992: Two Decades of Challenge*, New York: Chapman and Hall, 1992.

ENDA Tiers-Monde, 'Emergency Fight on Poverty, for Democracy and the Environment: The New Frontier.' Typescript, Dakar, Senegal (undated, but distributed at the 1992 Global Forum in Rio de Janeiro, Brazil).

Engel, Ronald J. *Sacred Dunes: The Struggle for Community in the Indiana Dunes.* Middletown, CT: Wesleyan University Press, 1983.

Environment Bulletin. Newsletter of the World Bank environment department, Washington, DC, 5(2) (Spring 1993):8 and 4(4) (Fall 1992):5.

Environment Canada, *Toxic Chemicals in the Great Lakes and Associated Effects*, 3 vols., Ottawa: Government of Canada, 1991.

Environmental Law Institute, *Protecting the Gulf of Aqaba: A Regional Environmental Challenge*, Washington, DC: Environmental Law Institute, 1993.

Environmental Protection Agency, *EPA Agreement Summary*, draft of proposed changes to the Great Lakes Agreement (original format); public, unreleased document, available in two forms: All Annex Format Draft and Index to Proposed Changes, Original Draft Form. 1987.

'The European Environmental Bureau: European Federation of Environmental NGOs'. Eight-page summary statement, Brussels, no date (distributed 1992).

'FAO Special Issue', *The Ecologist* 21(2) (March/April 1991).

Favre, David S. *International Trade in Endangered Species: A Guide to CITES*, Boston, MA: Martinus Nijhoff, 1989.

Finger, Matthias, 'The Military, the Nation State and the Environment', *The Ecologist* 21(5) (September/October 1991): 220–5.

Finger, Matthias, and Pascal Sciarini, *Homo politicus à la dérive? Enquête sur le rapport des Suisses à la politique*, Lausanne: Loisirs et Pedagogie, 1991.

Fitzgerald, Sarah, *International Wildlife Trade: Whose Business Is It?* Washington, DC: World Wildlife Fund, 1989.

Fri, Robert, 'Sustainable Development: Principles into Practice', *Resources* 102 (Winter, 1991): 1–3.

Friends of the Earth, Newsletter of Friends of the Earth-U.S., Washington, DC: 23 (January 1993), and 21 (November/December 1991).

George Gallup International Institute, 'The Health of the Planet Survey'. Quoted in 'Bush Out of Step, Poll Finds', *Terra Viva: The Independent Daily of the Earth Summit*, Rio de Janeiro, June 3, 1992.

Gibbins, John R., ed., *Contemporary Political Culture: Politics in the Postmodern Age*, Beverly Hills, CA: Sage Publications, 1989.

Gilbertson, Michael, 'Epidemics in Birds and Mammals Caused by Chemicals in the Great Lakes', in Marlene S. Evans, ed., *Toxic Contaminants and Ecosystem Health: A Great Lakes Focus*, vol. 21, New York: John Wiley, 1988: 133–52.

Gilbertson, Michael, 'Three Models of the Future', in George Modelski, ed., *Transnational Corporations and World Order*, San Francisco, CA: W.H. Freeman, 1979.

Glazer, Penina M., and Myron P. Glazer, 'Citizens' Crusade for a Safe Environment: Israel's Determined Few.' Paper presented at the eighth annual Israel Conference Day, University of Michigan, Ann Arbor, MI, January 31, 1993.

Glennon, Michael J. 'Has International Law Failed the Elephant?', *American Journal of International Law* 84 (1990): 1–43.

Goodland, Robert, Herman Daly, and Salah el-Serafy, eds, 'Environmentally Sustainable Economic Development: Building on Brundtland'. Environment Working paper no. 46, Washington, DC: World Bank, 1991.

Great Lakes Basin Commission, *Great Lakes Communicator.* 9 (1) 1978.

Great Lakes United, Water Quality Task Force, *Unfulfilled Promises: A Citizen's Review of the International Great Lakes Water Quality Agreement*, Buffalo, NY: Great Lakes United, 1987.

Greenpeace International. Various fact sheets. Expedition and field reports for 1987 to 1989. 'The Future of the Antarctic – Background for a Second UN Debate', October 22, 1984.

Grima, A. P., and R. J. Mason, 'Apples and Oranges: Toward a Critique of Public Participation in Great Lakes Decisions', *Canadian Water Resources Journal* 8(1) (1983): 22–50.

Gruner, Erich, and Hans-Peter Hertig, *Der Stimmbürger und die Neue Politik*, Berne: Haupt, 1983.

Guillen, Sergio, 'The Role of U.S.–Mexico Border Communities as Actors in Transboundary Environmental Policy.' Paper, University of Michigan, 1991.

Guimarães, Roberto P. *The Ecopolitics of Development in the Third World: Politics and Environment in Brazil*, Boulder, CO: Lynne Rienner Publishers, 1991.

Gurr, Ted R. *Why Men Rebel*, Princeton, NJ: Princeton University Press, 1970.

Haas, Peter M. *Saving the Mediterranean: The Politics of International Environmental Cooperation*, New York: Columbia University Press, 1990.

Haas, Peter M., ed, *Knowledge, Power, and International Policy Coordination* (special edition), *International Organization* 46, 1 (Winter 1992).

Habermas, Jürgen, *Legitimation Crisis*, Boston, MA: Beacon Press, 1973.

Habermas, Jürgen, 'New Social Movements', *Telos* 49 (1981): 33–7.

Hardin, Russell, *Collective Action*, Baltimore, MD: Johns Hopkins University Press, 1982.

Hartman, W. 'Historical Changes in the Major Fish Resources of the Great Lakes', in Marlene S. Evans, ed., *Toxic Contaminants and Ecosystem Health: A Great Lakes Focus*, vol. 21, New York: John Wiley, 1988: 103–31.

Hawkins, Ann, 'Contested Ground: International Environmentalism and Global Climate Change.' Paper presented at 1992 annual meeting of the International Studies Association, Atlanta, GA.

Hays, Samuel P. *Conservation and the Gospel of Efficiency: The Progressive Conservation Movement, 1890–1920*, Harvard Historical Monographs, Cambridge, MA: Harvard University Press, 1959.

Hecht, Susanna and Alexander Cockburn, *The Fate of the Forest: Developers, Destroyers and Defenders of the Amazon*, New York: Harper Perennial, 1990.

Hermann, Charles, 'Discussion of 'Turbulence in "World Politics"' by James Rosenau', *American Political Science Review* 85 (1991): 1081–3.

Hildyard, Nicholas, 'My Enemy's Enemies . . .', *The Ecologist* 23(2) (March/April 1993): 42–3.

Hirschman, A. O. *Exit, Voice, and Loyalty: Responses to Decline in Firms, Organizations, and States*, Cambridge, MA: Harvard University Press, 1970.

Illustrated World, 32(6) (February 1920): 926.

Hurrell, Andrew, and Kingsbury, Benedict, eds, *The International Politics of the Environment: Actors, Interests and Institutions*, Oxford: Oxford University Press, 1992.

Inglehart, Ronald, *The Silent Revolution: Changing Values and Political Styles Among Western Publics*, Princeton, NJ: Princeton University Press, 1977.

Inglehart, Ronald, *Culture Shift in Advanced Industrial Society*, Princeton, NJ: Princeton University Press, 1990.

Inskipp, Tim, and Sue Wells, *International Trade in Wildlife*, London: Earthscan Publications, 1979.

Interaction. Newsletter of Global Tomorrow Coalition, Washington, DC: 10(3) (Summer 1991): 4.

International Joint Commission, Great Lakes Water Quality Agreement, United States and Canada, Article 2, 'General Water Quality Objectives', 1972.

International Joint Commission. Fact Sheet: 1978 Great Lakes Water Quality Agreement: Ottawa: International Joint Commission, 1980.

International Joint Commission, *First Biennial Report under the Great Lakes Water Quality Agreement of 1978*, Windsor, Ontario: International Joint Commission, 1982.

International Joint Commission, *Remedial Action Plans For Areas of Concern*, informational brochure windsor, Ontario: International Joint Commission, 1985.

International Joint Commission, *Second Biennial Report*. Windsor, Ontario: International Joint Commission, December 31, 1984.

International Joint Commission, Great Lakes Diversions and Consumptive Uses, Windsor, Ontario: International Joint Commission, 1985.

International Joint Commission, 'International Joint Commission Activities', *Focus* 16 (1) (March/April 1991).

International Joint Commission, Great Lakes Water Quality Board, *Cleaning up Our Great Lakes: A Report on Toxic Substances in the Great Lakes Basin Ecosystem*, Windsor, Ontario: International Joint Commission, 1991.

International Joint Commission, Great Lakes Water Quality Board, *Review and Evaluation of the Great Lakes Remedial Action Plan (RAP) Program, 1991*, Windsor, Ontario: International Joint Commission, 1991.

International Joint Commission, Science Advisory Board, *1989 Report to the International Joint Commission*, Windsor, Ontario: International Joint Commission, 1989.

International Joint Commission, Virtual Elimination Task Force, *Persistent Toxic Substances: Virtually Eliminating Inputs to the Great Lakes*, Windsor, Ontario: International Joint Commission, 1991.

International Union for the Conservation of Nature/United Nations Environment Program/World Wildlife Fund, *The World Conservation Strategy: Living Resource Conservation for Sustainable Development*, Gland: International Union for the Conservation of Nature, 1980.

Ivory Trade Review Group, *The Ivory Trade and the Future of the African Elephant. Vol. 1: Summary and Conclusions*. Prepared for the Seventh CITES Conference of the Parties, Lausanne, October 1989.

Jackson, John, 'Citizen Involvement in the Review of the Great Lakes Water Quality Agreement.' Report to Environment Canada, 1991. Unpublished paper, available through the Great Lakes Research Consortium, Syracuse, NY.

Jackson, John, and Tim Eder, 'The Public's Role in Lake Management: The Experience in the Great Lakes', In *Socioeconomic Aspects of Lake/Reservoir Management*, International Lake Environment Foundation, United Nations Environment Programme, 1991: 31–46.

Jacobsen, Harold, *Networks of Interdependence: International Organizations and the Global Political System*, New York: Alfred Knopf, 1984.

Jancar-Webster, Barbara, 'Chaos as an Explanation of the Role of Environmental Groups in East Europe Politics'. Paper presented at the annual meeting of the International Studies Association, London, March 1989.

Japanese Working for a Better World: Grassroots Voices and Access Guide to Citizens' Groups in Japan. San Francisco, CA: HONNOKI USA, 1992.

Jenkins, Charles, 'Resource Mobilization Theory and the Study of Social Movements', *Annual Review of Sociology*, 9 (1983): 527–53.

Jhamtani, Hira, 'The Imperialism of Northern NGOs,' *Earth Island Journal*, San Francisco, CA: Earth Island Institute (7) (June 1992): 10.

Jolly, Allison, 'The Madagascar Challenge: Human Needs and Fragile Ecosystems', in H. Jeffrey Leonard, ed., *Environment and the Poor: Development Strategies for a Common Agenda*, New Brunswick: Transaction Books, 1989.

Keller, Hans, ed., *Who is Who in Service to the Earth*, Waynesville, NC: VisionLink Foundation, 1991.

Keohane, Robert O., and Joseph S. Nye, eds, *Transnational Relations and World Politics*, Cambridge, MA: Harvard University Press, 1972.

Kimball, L. *IIED Reports on Antarctica*, Washington, DC: International Institute for Environment and Development, 1984–1987.

Kimball, L. 'Political Opportunity Structures and Political Protest: Anti-Nuclear Movements in Four Democracies,' *British Journal of Political Science* 16 (1986): 57–85.

Kimball, L. *Special Report on the Antarctic Minerals Convention*, Washington, DC: International Institute for Environment and Development, February and July 1988.

Kimball, L. *Southern Exposure: Deciding Antarctica's Future*, Washington, DC: World Resources Institute, 1990.

Klandermans, Bert, and Sidney Tarrow, 'Mobilization into Social Movements: Synthesizing European and American Approaches,' in Bert Klandermans *et. al.*, eds, *International Social Movement Research*, vol.1., Greenwich, CT: JAI Press, 1988.

Korten, David, 'Community Management and Social Transformation,' in David Korten, ed., *Community Management. Asian Experiences and Perspectives*, West Hartford, CT: Kumarian Press, 1987: 319–28.

Korten, David 'Third Generation NGO Strategies: A Key to People-Centered Development', *World Development* 15 (Suppl. Autumn 1987): 145–59.

Korten, David *Getting to the 21st Century: Voluntary Action and the Global Agenda*, West Hartford, CT: Kumarian Press, 1990.

Korten, David, and Roger Klauss, eds, *People-Centered Development: Contributions toward Theory and Planning Framework*, West Hartford, CT: Kumarian Press, 1984.

Kosloff, Laura H., and Mark C. Trexler, 'The Convention on International Trade in Endangered Species: No Carrot, But Where's the Stick?', *Environmental Law Reporter* 17 (July 1987): 10222–36.

Krasner, Stephen D., ed, *International Regimes*, Ithaca, NY: Cornell University Press, 1983.

Kriesi, Hanspeter, 'New Social Movements and the New Class in the Netherlands', *American Journal of Sociology* 94(5) (1989): 1078–116.

Lancaster, John, 'Jay Hair's Environmental Impact: Playing Hardball at the National Wildlife Federation', *The Washington Post National Weekly Edition*, September 9–15, 1991: 12–13.

Leiss,William, ed., *Ecology vs. Politics in Canada*, Toronto: University of Toronto Press, 1979.

Leonard, Jeffrey H. *Environment and the Poor: Development Strategies for a Common Agenda*, New Brunswick, NJ: Transaction Books, 1989.

Lerner, Sally, 'A Study of Ontario Volunteer Environmental Stewardship Groups', Technical Paper no.6, Waterloo: Heritage Resource Center, 1992.

Lerner, Stephen, ed., *Earth Summit: Conversations with Architects of an Ecologically Sustainable Future*, Bolinas, CA: Common Knowledge Press, 1992.

Levine, Adeline Gordon, *Love Canal: Science, Politics, and People*, Lexington, MA: Lexington Books, 1982.

Lewis, John and contributors, *Strengthening the Poor: What Have We Learned?*, New Brunswick, NJ: Transaction Books, 1988.

Lindborg, Nancy, 'Nongovernmental Organizations: Their Past, Present, and Future Role in International Environmental Negotiations', in Lawrence E. Susskind, Eric Jay Dolin and J. William Breslin, eds, *International Environmental Treaty Making*, Cambridge, MA: Program on Negotiation at the Harvard Law School, 1992.

Lipschutz, Ronnie D. 'Global Change & Cooperation in the Implementation of International Environmental Agreements: From Theory to Practice'. Typescript, 1991.

Lipschutz, Ronnie D. 'From Here to Eternity: Environmental Time Frames and National Decisionmaking'. Paper presented at the 1991 annual meeting of the International Studies Association, Vancouver, BC, 1991.

Lipschutz, Ronnie D. 'Heteronomia: The Emergence of Global Civil Society'. Paper presented at the annual meeting of the International Studies Association, Atlanta, GA, March 31–April 4, 1992.

Lipschutz, Ronnie D. 'Reconstructing World Politics: The Emergence of Global Civil Society', *Millenium* 21(3) (Winter 1992): 389–420.

Lipschutz, Ronnie D. 'Learn of the Green World: Global Environmental Change, Global Civil Society and Social Learning'. Paper presented at the 34th Annual Conference of the International Studies Association, Acapulco, Mexico, March 23–7, 1993.

Lipsky, Michael, *Street-Level Bureaucracy: Dilemmas of the Individual in Public Services*, New York: Russell Sage Foundation, 1980.

Lohmann, Larry, 'Whose Common Future?', *The Ecologist* 20 (3): 82–84, editorial.

Long, Frederick J. 'The Nature Conservancy'. Case number S-PM-32, Stanford, CA: Graduate School of Business, Stanford University, 1992.

Lyster, Simon, *International Wildlife Law*, Cambridge: Grotius Publications, 1985.

McCormick, John, *Reclaiming Paradise: The Global Environmental Movement*, Bloomington: Indiana University Press, 1989.

McCormick, John, 'British Environmental Policy and the European Community'. Paper presented at 1991 annual meeting of the International Studies Association, Vancouver, BC.

MacNeill, Jim, Pieter Winsemius, and Taizo Yakushiji, *Beyond Interdependence: The Meshing of the World's Economy and the Earth's Ecology*, New York: Oxford University Press, 1991.

Manno, Jack, 'Citizen Participation and Consensus Building', in L. S. Bankert and R. W. Flint, eds, Great Lakes Monograph no. 1, *Environmental Dispute Resolution in the Great Lakes Region*, Buffalo, NY: Great Lakes Program, 1988: 65–6.

Manno, Jack, 'Federalist and Ecologist'. Working paper of the Great Lakes Research Consortium, 1991. Unpublished paper available from the Great Lakes Research Consortium, Syracuse, NY.

Marshall, George, 'Political Economy of Logging', *The Ecologist*, 20(5).

Mazur, Allan, and Jinling Lee, 'Sounding the Global Alarm: Environmental Issues in the National News', *Social Studies of Science*, 23 (1993) 681–720.

Meeker, Joy, 'Greenpeace', Paper, Syracuse University, NY, 1991.

Meidinger, Errol, *Community, Culture and Democracy in Administrative Regulation*. Working paper of Center for Law and Social Policy, Buffalo, NY, 1991.

Melucci, Alberto, *Nomads of the Present: Social Movements and Individual Needs in Contemporary Society*, London: Hutchinson, 1989.

Merchant, Carolyn, *Radical Ecology: The Search for a Livable World*, London: Routledge, Chapman and Hall, 1992.

Migdal, Joel S. *Strong Societies and Weak States: State–Society Relations and State Capabilities in the Third World*, Princeton, NJ: Princeton University Press,1988.

Milbrath, Lester, *Envisioning a Sustainable Society: Learning Our Way Out*. Albany, NY: State University of New York Press, 1989.

Milikin, Tom. 'Japan's Ivory Trade.' *TRAFFIC Bulletin* 7(3/4): 43.

Minear, L. 'The other missions of NGOs: Education and advocacy.' *World Development* 15 (suppl. Autumn 1987) 201–12.

Mingst, Karen. 'Implementing International Environmental Treaties: The Role of NGOs.' Paper presented at the annual meeting of the International Studies Association, Mexico, 1993.

Minister of Natural Resources and Tourism. 'Ivory Trade: The Zimbabwe Position', Department of National Parks and Wildlife Management, press conference, Harare,

Zimbabwe, September 22, 1989, 1–9.

Minister of Natural Resources and Tourism 'Current Developments Regarding the Proposed Ivory Trade Ban.' Press release, September 1989.

Mitchell, Barbara, and Jon Tinker, *Antarctica and its Resources*, London: Earthscan Publication: International Institute for Environment and Development, 1980.

Mitchell, Robert, Angela Mertig, and Riley Dunlap. 'Twenty Years of Environmental Mobilization: Trends among National Environmental Organizations'. *Society and Natural Resources* 4 (1991) 219–34.

Moltke, Konrad von, 'International Commissions and Implementation of International Environmental Law,' in John Carroll, ed., *International Environmental Diplomacy*, Cambridge: Cambridge University Press, 1988.

Müller-Rommel, Ferdinand, ed., *New Politics in Western Europe: The Rise and Success of Green Parties and Alternative Lists*, Boulder, CO: Westview Press, 1989.

Muntemba, Shimwaayi, 'Research for Sustainable Development: The Role of NGOs', *Development* 2(3) (1989): 65–7.

Munton, Donald, 'Paradoxes and Prospects', in Robert Spencer, John Kirtin, Kim Richard Nossal, eds, *The International Joint Commission Seventy Years On*, Toronto: Center for International Studies, University of Toronto, 1981: 60–97.

Munton, Donald, 'Toward a More Accountable Process: The Royal Society-National Research Council Report', in Lynton K. Caldwell, ed., *Perspectives on Ecosystem Management for the Great Lakes*, Albany: State University of New York Press, 1985: 299–317.

Nachowitz, Todd, ed., *An Alternative Directory of Nongovernmental Organizations in South Asia*, rev. edn, Syracuse, NY: Maxwell School of Citizenship and Public Affairs, Syracuse University; distributed by Foreign and Comparative Studies, South Asian Studies, and Fourth World Press, 1990.

Nadelmann, Ethan A. 'Global Prohibition Regimes: The Evolution of Norms in International Society', *International Organization* 44(4): 479–526.

National Research Council of the United States and the Royal Society of Canada, *The Great Lakes Water Quality Agreement: An Evolving Instrument for Ecosystem Management*, Washington, DC: National Academy Press, 1985.

Nerfin, Marc, 'Neither Prince nor Merchant – An Introduction to the Third System.' IFDA dossier 56 (1986): 3–29.

Oberschall, Antony, *Social Conflict and Social Movements*, Englewood Cliffs, NJ: Prentice-Hall, 1973.

O'Connell, Michael A., and Michael Sutton, in cooperation with TRAFFIC–US, *The Effects of Trade Moratoria on International Commerce in African Elephant Ivory: A Preliminary Report*, Washington, DC: World Wildlife Fund and The Conservation Foundation, June 1990.

Offe, Claus, *Contradictions of the Welfare State*, Cambridge, MA: MIT Press, 1984.

Offe, Claus, *Disorganized Capitalism: Contemporary Transformations of Work and Politics*, Cambridge, MA: MIT Press, 1985.

Olsen, Jennifer, 'Italian Hazardous Waste in Koko, Nigeria', Paper, University of Michigan, Ann Arbor, 1993.

Olson, Mancur, *The Logic of Collective Action: Public Goods and the Theory of Groups*. Cambridge, MA: Harvard University Press, 1965.

Opp, Karl-Dieter, 'Konventionnelle und Unkonventionnelle Politische Partizipation', *Zeitschrift für Soziologie* 14(4): 282–96.

O'Riordan, Timothy, 'The Politics of Sustainability', in *Sustainable Environmental Management. Principles and Practice*, Boulder, CO: Westview Press, 1988: 29–50.

Osherenko, Gail, and Oran R. Young, *The Age of the Arctic: Hot Conflicts and Cold Realities*, Cambridge: Cambridge University Press, 1989.

Ostertag, Bob, 'Greenpeace Takes Over the World', *Mother Jones* (March/April 1991): 85.

Ostrom, Elinor, *Governing the Commons: The Evolution of Institutions for Collective Action*, Cambridge: Cambridge University Press, 1990.

Paehlke, Robert C. *Environmentalism and the Future of Progressive Politics*. New Haven, CT: Yale University Press, 1989.

Parikh, Jagdish, 'Indigenous-Tribal Forest Peoples' Alliance Aims to Halt "Ecocide and Ethnocide"', March 26, 1992, by NGONET on Econet conference, cdp:en.unced.gener.

Patel, J., and Susan May, eds, *Antarctica, The Scientists' Case for a World Park*, London: Greenpeace International, 1991.

Patterson, Alan, 'Debt-for-Nature', *Environment* 32(10): 5–13, 31–2.

Pearce, Fred, *Green Warriors: The People and the Politics Behind the Environmental Revolution*, London: Bodley Head, 1991.

Peluso, Nancy Lee, 'Coercing Conservation: The Politics of State Resource Control'. Paper presented at 1992 annual meeting of the International Studies Association, Atlanta, GA.

Penna, David R. 'Regulation of the Environment in Traditional Society as a Basis for the Right to a Satisfactory Environment'. Paper presented at the annual meeting of the International Studies Association, Mexico, 1993.

'Philippine Environmentalists Arrested for Efforts to Save Rainforest', *Haribon Update*, newsletter of the Haribon Foundation, 6(2) (March–April 1991): 1, 4.

Polar Research Board and the Commission on Physical Sciences, eds, *Antarctic Treaty System: An Assessment*. Proceedings of a workshop at Beardsmore, South Field Camp, Antarctica, January 7–13, 1985. Washington, DC: National Academy Press, 1986.

Pollack, Andrew, 'Agency Takes Step Toward Banning Whaling in Southern Ocean', *New York Times*, May 15, 1993: 2.

Porter, Gareth, and Janet Welsh Brown, *Global Environmental Politics*, Boulder, CO: Westview Press, 1991.

Princen, Thomas, *Intermediaries in International Conflict*, Princeton, NJ: Princeton University Press, 1992.

Princen, Thomas, 'The Role of Non-Governmental Organizations in Fostering International Environmental Governance', in Environmental Law Institute, *Protecting the Gulf of Aqaba: A Regional Environmental Challenge*, Washington, DC: Environmental Law Institute, 1993.

Princen, Thomas, 'From Timber to Forest: The Evolution of the Tropical Timber Trade Regime', Typescript 1993, University of Michigan, Ann Arbor, Michigan.

Princen, Thomas, 'Ivory, Conservation and Transnational Environmental Coalitions', in Thomas Risse-Kappen, ed., *Bringing Transnational Relations Back In*, (forthcoming).

Program for Zero Discharge, '*A Prescription for Healthy Great Lakes,*' Washington, DC: National Wildlife Federation and Canadian Institute for Environmental Law and Policy, 1991.

Pryde, Philip R. *Environmental Management in the Soviet Union*, Cambridge: Cambridge University Press, 1991.

Quigg, P. *Antarctica: The Continuing Experiment*, New York: Foreign Policy Association, 1985.

Raiffa, Howard, *The Art and Science of Negotiation*, Cambridge, MA: Harvard University Press, 1982.

Raschke, Joachim, *Soziale Bewegungen*, Frankfurt: Campus, 1985.

Redclift, Michael, *Sustainable Development: Exploring the Contradictions*, London: Methuen, 1987.

Reese, William, 'The Ecology of Sustainable Development', *The Ecologist* 20 (1): 18–23.

'Reporters' Notebook', *Earth Summit Times,* official newspaper of record of UNCED New York/Rio de Janeiro, June 5, 1992.

Richards, John F. and Richard Tucker, eds, *World Deforestation in the Twentieth Century,* Durham, NC: Duke University Press, 1988.

Riker, James V. 'Linking Development from Below to the International Environmental Movement: Sustainable Development and State–NGO Relations in Indonesia'. Paper presented at annual meeting of Northwest Regional Consortium for Southeast Asian Studies on 'Development, Environment, Community and the Role of the State', University of British Columbia, Vancouver, 16–18 October 1992.

Risse-Kappen, Thomas, ed., *Bringing Transnational Relations Back In,* Cambridge: Cambridge University Press (forthcoming).

Rosenau, James, *Turbulence in World Politics: A Theory of Change and Continuity,* Princeton, NJ: Princeton University Press, 1990.

Runge, C. Ford, 'Common Property and Collective Action in Economic Development', in Daniel W. Bromley, ed., *Making the Commons Work: Theory, Practice, and Policy,* San Francisco, CA: Institute for Contemporary Studies Press, 1992: 17–39.

Sabatier, Paul A. 'Top-Down and Bottom-Up Approaches to Implementation Research: a Critical Analysis and Suggested Synthesis', *Journal of Public Policy* 6 (1986): 21–48.

Sands, Philippe J., and Albert P. Bedecarre, 'Convention on International Trade in Endangered Species: The Role of Public Interest Non-governmental Organizations in Ensuring the Effective Enforcement of the Ivory Trade Ban', *Boston College Environmental Affairs Law Review* 17 (Summer, 1990): 799–822.

Schelling, Thomas C. *The Strategy of Conflict,* Cambridge, MA: Harvard University Press, 1960; reprint 1980.

Schneider, Bertrand, *The Barefoot Revolution: A Report to the Club of Rome,* London: Intermediate Technology Publications, 1988.

Schneider, Stephen, 'The Science/Policy Interface: How Much Knowledge Should Precede Action?', Newsletter of the Project for the Integrated Study of Global Change, University of Michigan, *Global Change* 1(2)(1991): 1–4.

Sewell, Derrick and J. T. Coppock, *Public Participation in Planning,* New York: John Wiley & Sons, 1977.

Shapley, D. *The Seventh Continent: Antarctica in a Resource Age,* Washington, DC: Resources for the Future, 1985.

Shiva, Vandana, *Staying Alive: Women, Ecology, and Development,* London: Zed Books, 1989.

Sierra Club, 'Great Lakes Washington Report' 1(10).

Smelser, Nelson J. *Theory of Collective Behaviour,* New York: Macmillan, 1963.

Sneddon, Chris, and Alma Lowry, 'Friends of the Earth International: The Role of an International Network in Global Environmental Politics'. Paper, School of Natural Resources and Environment, University of Michigan, April 1992.

Soroos, Marvin S., and Elena N. Nikitina, 'The World Meteorological Organization as a Purveyor of Global Public Goods'. Paper presented at the annual meeting of the International Studies Association, Mexico, 1993.

Stairs, Kevin, and Peter Taylor, 'Non-Governmental Organizations and Legal Protection of the Oceans: A Case Study' in Andrew Hurrell and Benedict Kingsbury, eds, *The International Politics of the Environment: Actors, Interests, and Institutions,* Oxford: Oxford University Press, 1992: 110–41.

Starke, Linda, *Signs of Hope: Working towards Our Common Future,* Oxford: Oxford University Press, 1990.

Swinehart, Carol Y. 'A Review of Public Participation in the Great Lakes Water Quality Agreement', in David H. Hickcox, ed., *Great Lakes: Living with North America's Inland Waters*. Proceedings of the American Water Resources Association Symposium, 1988.

Tandon, Yash, 'Foreign NGOs, Uses and Abuses: An African Perspective'. IFDA Dossier, no. 81 (April/June 1991): 68–78.

Tarrow, Sidney, 'Social Movements, Resource Mobilization and Reform during Cycles of Protest: A Bibliographical and Critical Essay'. Center for International Studies, Occasional paper no. 15. Ithaca, NY: Cornell University, 1982.

Teclaff, Ludwig, and Eileen Teclaff, 'Transboundary Toxic Pollution and the Drainage Basin Concept', in Albert Utton and Ludwig Teclaff, eds, *Transboundary Resources Law*, Boulder, CO: Westview Press, 1987.

Telos, 'Social Movements', *Telos* 49 (1981).

Thomas, Caroline, *The Environment in International Relations*, London: The Royal Institute of International Affairs, 1992.

Thomas, Urs, 'The UN Environment Programme–Constraints and Strategy in the Context of 1992'. Paper presented at the 31st Annual Convention of ISA, Washington, DC, April 10–14, 1990.

Thornton, Allan, 'A Ban on Ivory to Save the Elephant?', *New Scientist* (September 30, 1990): 43.

Tickell, Oliver, and Nicholas Hildyard, 'Green Dollars, Green Menace', *The Ecologist* 22(3) (May/June 1992): 82–3.

Tilly, Charles, *From Mobilization to Revolution*, Reading, MA: Addison and Wesley, 1978.

Tolba, Mostafa, Osama El-Kholy, E. El-Hinnawi, M.W. Holdgate, D.F. McMichael, and R.E. Munn, eds, *The World Environment 1972–1992: Two Decades of Challenge*, London: Chapman and Hall, 1992.

Touraine, Alain, *The Self-Production of Society*, Chicago, IL: University of Chicago Press, 1978.

Touraine, Alain, *The Post-Industrial Society*, New York: Random House, 1981.

Touraine, Alain, *The Voice and the Eye: An Analysis of Social Movements*, Cambridge: Cambridge University Press, 1981.

Touraine, Alain, *Anti-Nuclear Protest*, New York: Cambridge University Press, 1983.

Touraine, Alain, 'An Introduction to the Study of Social Movements', *Social Research* 52 (4) 1985: 749–87.

Touraine, Alain, *The Return of the Actor*, Minneapolis, MN: University of Minnesota Press, 1988.

TRAFFIC-USA newsletters: 10(1) (March 1990); 11(2) (January 1992).

Trzyna, Thaddeus C., and Roberta Childers, eds, *World Directory of Environmental Organizations*, 4th edn, Sacramento, CA: California Institute of Public Affairs, 1992.

UNCED Secretariat, *In Our Hands*, Earth Summit '92, Geneva, 1991.

UNESCO, *Use and Conservation of the Biosphere*, Proceedings of the Biosphere Conference, Paris: UNESCO, 1970.

Union of International Associations, *Yearbook of International Organizations 1988/89*. Munich: Saur Verlag, 1988.

United Nations Association, *Uniting Nations for the Earth: An Environmental Agenda for the World Community*, New York: UNA-USA Publications, 1990.

'US Agrees to Protect Minerals in Antarctica', *New York Times*, July 6, 1991.

'US Backdown on Antarctic Mining Ban Condemned', *The Melbourne Age*, June 24, 1991.

US Citizens Network on UNCED Newsletter, no. 13, April 1992, San Francisco, CA.

US Citizens Network newsletter, July 2, 1992.

US Department of State, telegram from American Embassy, Gaborone, Botswana, to Department of State, February 12, 1990 [R 121326Z Feb 90].

US Department of State, telegram from American Embassy, Abu Dhabi, UAE, to Department of State, May 16, 1990.

US Fish and Wildlife Service, Department of Interior, News Release, February 16, 1990.

US Fish and Wildlife Service, 'Moratorium on Importation of Raw and Worked Ivory From all Ivory Producing and Intermediary Nations', Notices, *Federal Register*, 54 (110) Friday, June 9, 1989: 24758–61.

'US on Thin Ice in Antarctica', *The Australian*, June 24, 1991.

'US Opposes Antarctic Mining Ban Now', *New York Times*, June 23, 1991.

'US Raises Objections to Antarctic Pact', *New York Times*, June 18, 1991.

'US Weighs Its Position on Drilling in Antarctica', *Los Angeles Times*, June 11, 1991.

Utton, Albert and Ludwig Teclaff, eds, *Transboundary Resources Law*, Boulder, CO: Westview Press, 1987.

Vicuna, F. *Antarctic Resource Policy*, Cambridge: Cambridge University Press, 1983.

Ward, Barbara and René Dubos, *Only One Earth: The Care and Maintenance of a Small Planet*, Harmondsworth: Penguin, 1972.

Weller, Phil, *Fresh Water Seas: Saving the Great Lakes*, Toronto: Between the Lines, 1990.

Westermeyer, William and Christopher Joyner, 'Negotiating a Minerals Regime for Antarctica', Pew Program in Case Teaching and Writing in International Affairs, Graduate School of Public and International Affairs, Case no. 134. Pittsburgh: Pittsburgh University Press, 1989: 1–34.

'What's the Hurry in Antarctica?', *New York Times*, June 6, 1991.

Wind, Adrian de, 'Alar's Gone, Little Thanks to the Government', *New York Times*, July 30, 1991.

Winham, Gilbert R. 'Negotiation as a Management Process', *World Politics* 30 (1977): 87–114.

Worster, Donald, ed., *The Ends of the Earth: Perspectives on Modern Environmental History*, Cambridge: Cambridge University Press, 1988.

World Resources Institute in collaboration with the United Nations Envrionment Programme and the United Nations Development Programme, *World Resources 1990–91: A Guide to the Global Environment*, New York: Oxford University Press, 1990.

World Resources Institute in collaboration with the United Nations Environment Programme and the United Nations Development Programme, *World Resources 1992–93: A Guide to the Global Environment*, New York: Oxford University Press, 1992.

World Wildlife Fund, 'A Program to Save the African Elephant', *World Wildlife Fund Letter*, no. 2: 1–12, Washington DC, 1989.

World Wildlife Fund Factsheet, 'Monitoring Wildlife Trade – The TRAFFIC Network', March 1989.

World Wildlife Fund, 1991 Annual Report.

Worldwise and Friends of the Earth, *International Directory of Non-Governmental Organizations*, Sacramento, CA: Worldwise and Friends of the Earth, 1992.

Young, Nigel, *An Infantile Disorder? The Crisis and Decline of the New Left*, Boulder, CO: Westview Press, 1977.

Young, Oran R. *International Cooperation: Building Regimes for Natural Resources and the Environment*, Ithaca, NY: Cornell University Press, 1989.

Young, Oran R., and Gail Osherenko, (eds), *Polar Politics: Creating International Envrionmental Regimes*, Ithaca NY: Cornell University Press, 1993.

Young, Oran R., George J. Demko, and Kilaparti Ramakrishna, 'Global Environmental Change and International Governance'. Summary and recommendations of a conference held at Dartmouth College, Hanover, NH, June 1991.

Zald, Mayer N., and John D. McCarthy, *Social Movements in an Organizational Society*, New Brunswick: Transaction Books, 1987.

Zald, Mayer N., and John D. McCarthy, eds, *The Dynamics of Social Movements: Resource Mobilization, Social Control and Tactics*, Cambridge, MA: Winthrop, 1979.

Zartman, I. William, 'Prenegotiation: Phases and Functions', in Janet Gross Stein, ed., *Getting to the Table: The Processes of International Prenegotiation*, Baltimore, MD: Johns Hopkins University Press, 1989.

Index